# World Internet Development Report 2020

Chinese Academy of Cyberspace Studies

# World Internet Development Report 2020

## Blue Book for World Internet Conference

中国工信出版集团 電子工業出版社 PUBLISHING HOUSE OF ELECTRONICS INDUSTRY http://www.phei.com.cn Springer

Chinese Academy of Cyberspace Studies
Chinese Academy of Cyberspace Studies
Beijing, China

ISBN 978-981-16-9390-8 ISBN 978-981-16-9388-5 (eBook)
https://doi.org/10.1007/978-981-16-9388-5

Jointly published with Publishing House of Electronics Industry
The print edition is not for sale in China (Mainland). Customers from China (Mainland) please order the print book from: Publishing House of Electronics Industry.



This Springer imprint is published by the registered company Springer Nature Singapore Pte Ltd.
The registered company address is: 152 Beach Road, #21-01/04 Gateway East, Singapore 189721, Singapore

# Preface

The world today is marked by changes unseen in a century. New technologies like 5G, artificial intelligence, big data and blockchain play a leading role in a deep-level scientific and technological revolution and industrial revolution that humans must face. Since 2020, information technology has evolved vigorously and the digital economy has developed rapidly. As a result, the interests and fate of various countries are tightly tied.

The sudden outbreak of COVID-19 means a severe test for the world. COVID-19 tells us that humans live in a global village in which we are interconnected and interdependent. Closely linked, all countries share a common fate. No country can reap the benefits from others' tribulation and turbulence. As a Chinese idiom suggests, if a country watches a fire from the opposite bank of the river or does harm to a neighbor, it will draw ruin upon itself. Humans must forge the consciousness of a community with a shared future featuring interconnection and interdependence, break the limitation of cliquism and zero-sum game, build the ideas of holism and win-win cooperation, spurn ideological debate, avert the trap of clash of civilizations, and respect the developmental path and model that a country autonomously chooses. In this way, the idea of "diversity of the world" sustainably supports the social progress of humans and naturally displays the diversity of human civilization.

Against this backdrop, we plan and compile *World Internet Development Report 2020* (hereinafter referred to as "The Report"). Using the index and indicator system in the last year, we evaluate the world Internet development, summarize the achievements and experience in the construction of national network information infrastructure, analyze the latest achievements and practical use of the world information technology, and encapsulate the new trend and success in the development of the digital economy in various countries. Noticeably, we sum up the practice and experience concerning how major countries cope with COVID-19 by using the Internet, for reference only. "The Report" represents the technological innovation and practical use of the world Internet development in many countries in 2020, covering eight aspects, i.e. information infrastructure, information technology, digital economy, e-government, Internet media, cybersecurity, cyberlaw and international governance. Our aim is to promote the interconnectivity and sharing and co-governance in the world Internet development, so that the Internet provides strong support for overcoming COVID-19 and resuming work and production in all countries and benefits people of all countries in a better and faster way.

"The Report" crystallizes the important achievements made by the Chinese academic circle that hopes to provide theoretical thought and experience for the world Internet development and governance. In the future, we continue to follow the trend and advancement of the world Internet development, offer our analyses and insights, and contribute our wisdom and experience to quickening the building of a community with a shared future in cyberspace.

Beijing, China
October 2020

Chinese Academy of Cyberspace Studies

# Contents

**1 Overview** .... 1
1.1 Overall Trends in World Internet Development in 2020 .... 1
1.1.1 COVID-19 Pandemic Hits Global Economic and Social Development, and Digital Economy is Viewed as a New Engine for the Recovery of Global Economy .... 2
1.1.2 Internet Media Achieves Diversified Development, and All Countries Continuously Strengthen Cyber Content Governance .... 3
1.1.3 All Countries in the World Actively Deploy New-Type Infrastructure and Accelerate the Construction of Information Infrastructure .... 3
1.1.4 Innovation in the Emerging Internet Technology Shows Vitality, and Various Countries Strive to Bridge the Digital Divide .... 4
1.1.5 Cybersecurity Threats Become More Prominent, and All Countries Constantly Strengthen Cybersecurity Protection .... 5
1.1.6 Internet Facilitates the Exchange and Cooperation Among Countries in the World, and International Cyberspace Governance Needs to Be Explored and Promoted .... 6
1.2 Internet Development in Representative Countries .... 7
1.2.1 Index Construction .... 7
1.2.2 GIDI System .... 7
1.2.3 Analysis of Results .... 14
1.3 Further Information on Selected Countries .... 26
1.3.1 The United States Maintains the Global Leading Role in Comprehensive Strength of Internet .... 28
1.3.2 China's Internet Achieves Stable and Rapid Development on the Whole .... 30

1.3.3 The United Kingdom Achieves Balanced Overall Strength in the Internet ..... 32
1.3.4 Singapore Strives to Build the World's Leading Digital Nation ..... 35
1.3.5 Sweden Displays Excellent Capacity in Internet Innovation ..... 36
1.3.6 Canada Possesses Prominent Advantages in Digital Governance and Cybersecurity ..... 38
1.3.7 Japan Accelerates the Development of Mobile Internet and Digital Economy ..... 39
1.3.8 Australia Takes the Lead in the Internet Universalization Rate in the World ..... 42
1.3.9 Italy Maintains High Level in the Internet R&D Expenditure ..... 43
1.4 Trends of World Internet Development ..... 46
1.4.1 Under the Circumstance of the COVID-19 Pandemic, the Degree of Dependence on Internet Increases, and All Countries Constantly Strengthen Information Infrastructure Construction ..... 47
1.4.2 The COVID-19 Pandemic Stimulates the Demand for Cloud-Terminal Coordination and Intelligentization, and Internet-Related Industries Face a New Development Opportunity ..... 47
1.4.3 Global Digital Economy Faces Great Uncertainty, and Global Internet Technological Ecology Displays a Diverse Trend ..... 48
1.4.4 The Imbalanced Digital Development in Various Countries Probably Further Widens the World Digital Divide, and the Demand for Global Sharing of Digital Advantages Becomes More Urgent ..... 48
1.4.5 Internet Omni-Media Displays a Distinct Trend, and Content Delivery Develops Toward a Refined State ..... 49
1.4.6 The Militarization and Competition of Cyberspace Intensify, and the World's Cybersecurity Situation Becomes More Serious ..... 50
1.4.7 Global Cyberspace Landscape Accelerates Its Change, and International Internet Governance Model Faces Adjustment ..... 50
1.5 Joining Hands to Build a Community with a Shared Future in Cyberspace ..... 51

**2 The Construction of World Information Infrastructure** ........ 53
2.1 Outline ........ 53
2.2 The Evolution and Upgrading of Basic Network Speed Up ........ 54
2.2.1 Global 5G Network Develops Rapidly ........ 54
2.2.2 4G Network Achieves Deep Coverage ........ 56
2.2.3 The Construction of High-Speed Broadband Network Accelerates ........ 56
2.2.4 International Network Construction Meets a Crucial Window Period ........ 58
2.2.5 Competition on the Construction of Spatial Information Infrastructure Turns White-Hot ........ 60
2.3 Application-Oriented Infrastructure Develops Steadily ........ 62
2.3.1 The Number of Global Data Centers Grows Stably ........ 62
2.3.2 The Infrastructure of Cloud Computing and Edge Computing Displays Enormous Potential in Development ........ 63
2.3.3 AI Platform Faces a Period of Rapid Development ........ 64
2.3.4 Blockchain Encounters a Period of Outbreak ........ 65
2.3.5 Domain Name Market and IPv6-Construction Advance Solidly ........ 65
2.4 The Global Deployment of New Facilities Speeds Up ........ 67
2.4.1 The Deployment of Global IoT Facilities Quickens ........ 68
2.4.2 The Construction of Global IoT Platform Accelerates ........ 68
2.4.3 The Construction of Industrial Internet Makes Significant Progress ........ 70

**3 The Development of World Information Technology** ........ 75
3.1 Outline ........ 75
3.2 Basic Technologies ........ 76
3.2.1 High-Performance Computing (HPC) Initially Achieves the Exascale Level ........ 76
3.2.2 Chip Technology Faces Technological Revolution ........ 79
3.2.3 The Integration and Intelligentization of Software Technology Develop Faster ........ 82
3.3 Cutting-Edge Technologies ........ 86
3.3.1 AI Keeps High-Speed Development ........ 86
3.3.2 Key Innovative Achievements Continuously Emerge in Quantum Information ........ 89
3.3.3 Biocomputing and Biostoring Achieve Key Breakthroughs ........ 92
3.3.4 Brain-Computer Interface (BCI) Progressively Develops from an Idea into Prototype Design ........ 94
3.4 New Technologies and New Applications ........ 96
3.4.1 WITMED Realizes Rapid Development ........ 96
3.4.2 Intelligent Traffic App Gradually Spreads ........ 98

3.4.3 Smart Home Develops in a Diverse Way and Competition Among Enterprises Intensifies ......... 99
3.4.4 The Industrial Upgrading of Intelligent Manufacturing Accelerates ......................... 100
References ......................................................... 101

**4 The Development of World Digital Economy** ..................... 103
4.1 Outline ....................................................... 103
4.2 The Developmental Trend of World Digital Economy ............ 104
4.2.1 The Developmental Strategy Becomes More Targeted .... 104
4.2.2 The Developmental Pattern Basically Maintains Stable ............................................... 106
4.2.3 Internet Enterprises Rise Rapidly ..................... 107
4.2.4 Investment Markets in Chinese and American Digital Economy Thrive ............................ 108
4.2.5 Global Digital Trade: Development & Potential Danger ............................................... 108
4.2.6 Digital Economy Contributes to Global Fight Against the COVID-19 Pandemic ..................... 110
4.3 Digital Industrialization Grows Steadily ....................... 110
4.3.1 Basic Telecommunications Progress Stably ............. 110
4.3.2 Electronics and Information Manufacturing Industry Displays Great Potential ............................ 111
4.3.3 Software and Information Technology Service Industry Advances Smoothly ......................... 113
4.3.4 Internet Information Content Service Industry Expands Fast .......................................... 116
4.4 Industrial Digitization Advances Deeply ...................... 117
4.4.1 Agricultural Digitization Develops Firmly .............. 118
4.4.2 Manufacturing Digitization Continues to Deepen ........ 119
4.4.3 Service Industrial Digitization Realizes Transformation and Upgrading ........................ 120
4.5 The Standardization of Fintech Parallels Its Development ........ 121
4.5.1 Issuing Digital Currency is Put on the Agenda ........... 122
4.5.2 Digital Bank Becomes a Popular Developmental Trend ............................................... 122
4.5.3 Emerging Economies Show Vigor in Development ....... 122
4.5.4 The Supervision of Digital Finance Gradually Improves ............................................ 123
4.6 E-Commerce Keeps Expanding ................................ 123
4.6.1 The Sales of E-Commerce Continually Rise ............ 123
4.6.2 Global Market Pattern Basically Remains Steady ........ 125

**5 The Development of World E-Government** ........ 129
5.1 Outline ........ 129
5.2 The Leading Exploration of World E-Government ........ 129
5.2.1 The Innovation on Technology and Application Emerges in E-Government ........ 130
5.2.2 E-Government Stimulates Innovation in Service Model ........ 131
5.2.3 E-Government Triggers Coordinated Innovation in Governance ........ 133
5.3 Practical Evaluation of World E-Government ........ 134
5.3.1 Comprehensive Evaluation of E-Government ........ 134
5.3.2 Infrastructure of E-Government ........ 135
5.3.3 Open Level of Governmental Data ........ 138
5.3.4 Level of Online Service ........ 141
5.3.5 Level of Digital Literacy ........ 142
5.4 Annual Hot Words: E-Government Under the Outbreak of COVID-19 ........ 143
5.4.1 World Health Organization Applied E-Government to the Prevention and Control of COVID-19 ........ 144
5.4.2 Several European Countries Launched "Hackathon" Competition to Fight Against COVID-19 in a Coordinated Way ........ 145
5.4.3 Germany and Italy Evened Out Personal Travel Tracking and Privacy Protection ........ 147
5.4.4 Singapore Carried Out Data-Based Meticulous Management in the Fight Against COVID-19 ........ 148
5.4.5 South Korea Positively Developed Mobile Apps in the Outbreak of COVID-19 and Provided Effective Services ........ 149
References ........ 150

**6 The Development of World Internet Media** ........ 151
6.1 Outline ........ 151
6.2 The Developmental Pattern of World Internet Media ........ 152
6.2.1 Digital Media Presents Diverse Developmental Trends ........ 152
6.2.2 The Use Cases of Cutting-Edge Technologies Become Increasingly Diverse ........ 156
6.3 The Clash and Cooperation Among World Internet Media Co-Exist ........ 158
6.3.1 "Infodemic" Comes to Rise in the Outbreak of COVID-19 ........ 159

6.3.2 COVID-19 Gives Impetus to Global Digital Cooperation Course ... 160
6.3.3 Divergences on the Values of World Internet Media in the Outbreak of COVID-19 ... 162
6.4 Hot Topics Among World Internet Media ... 163
6.4.1 Computational Propaganda Affects the Ecosystem of World Internet (Media) ... 163
6.4.2 Digital Omni-Media Displays a Prominent Trend ... 164
6.4.3 Transnational Internet (Media) Cooperation Continually Deepens ... 165
6.4.4 The Content Delivery of Internet Media Advances Toward Meticulousness ... 166
References ... 167

**7 The Development of World Cybersecurity** ... 169
7.1 Outline ... 169
7.2 World Cybersecurity Develops into a New Stage ... 170
7.2.1 The Evolution of Cyberspace Results in New Changes in Cybersecurity ... 170
7.2.2 The Game Among Major Powers (Competition & Cooperation) Poses New Challenges to Cybersecurity ... 171
7.2.3 COVID-19 Intensifies the Oscillation of Cybersecurity ... 171
7.3 World Cybersecurity Threats Show New Features ... 173
7.3.1 Trends of Threats in Attacks on Cybersecurity ... 173
7.3.2 Trends of Threats to Major Platforms & Fields and Cybersecurity ... 177
7.4 Various Countries in the World Have Made New Achievements in Formulating Policies on Cybersecurity ... 179
7.4.1 Establishing Strategies on Cybersecurity: A Survey ... 179
7.4.2 The Protection of Data Security Has Been Constantly Enhanced ... 180
7.4.3 The Protection of Critical Information Infrastructure Continues to Be Strengthened ... 181
7.4.4 Policies on Technological Field Have Been Actively Enacted ... 182
7.5 New Trends in the Technological Development of World Cybersecurity ... 182
7.5.1 Zero Trust Architecture Multiplies ... 183
7.5.2 Human Factors Play a Prominent Role in Cybersecurity ... 183
7.5.3 Intelligent Means Are Used to Solve Problems on Cybersecurity More Widely ... 183

7.6 Cybersecurity Industry Develops Sustainably .................. 184
7.6.1 World Cybersecurity Industrial Scale Grows Steadily .... 184
7.6.2 Cybersecurity Industry Advances Progressively ......... 184
7.6.3 The Construction of Cybersecurity Industrial Ecosystem ........................................... 186
7.7 New Demands for Talent Training in Cybersecurity ............. 187
7.7.1 The Lack of Talents and Technologies in World Cybersecurity Remains Grim Now .................... 187
7.7.2 New Measures Taken to Strengthen the Talent Training in Cybersecurity ............................ 190

**8 The Construction of World Cyberlaws** .......................... 191
8.1 Outline ........................................................ 191
8.2 The Legislative Work of Personal-Information Protection Further Advances, with the Formulation of Special Rules for Industrial Data Explored ................................ 192
8.2.1 Legislative Object Develops Deeply, with Specific Rules Continually Improved ......................... 192
8.2.2 Basic Principles for Information Protection Maintain Intact, with Exceptional Cases Allowed for Epidemic-Related Data ........................... 194
8.2.3 Data Leakage Occurs Frequently, with Administrative Penalty Tightened ................. 196
8.3 Strengthening the Regulation of Cyber Platforms and Purifying the Environment of Cyber Platforms .............. 197
8.3.1 Standardizing the Competitive Means in Cyber Platforms and Maintaining the Free-Competition Order in Cyber Platforms ............................ 197
8.3.2 Protecting the Space for Consumers' Choice and Clarifying the Rules for Releasing Information on Cyber Platforms ................................. 198
8.3.3 Enhancing the Supervision of Cyber Content and Purifying the Environment of Cyber Platforms ....... 199
8.4 Promoting the Legislative Course of Cybersecurity and Improving the Guarantee System of Cybersecurity ........... 200
8.4.1 Establishing Cybersecurity Departments and Admitting Social Participants .................... 200
8.4.2 Improving Cybersecurity Systems and Attaching Great Attention to Supply-Chain Security ............. 201
8.4.3 Long Arm Jurisdiction is Frequently Used, Which Provokes International Concern ...................... 202

8.5 New Technologies and New Business Forms Accelerate Iteration, the Risk-Prevention Mechanism Gradually Matures ..... 203
8.5.1 The Legislation of Digital Payment Incessantly Improves, the Legal Supervision of Blockchain Gradually Plays a Dominant Role ..... 203
8.5.2 Drawing the Blueprint of AI Development, Constructing the Moral Norms for AI ..... 204
8.5.3 Internet of Things (IoT) Industry Opens up New Situation, Rules for Security Guarantee Initially Take Shape ..... 205

**9 International Cyberspace Governance** ..... 209
9.1 Outline ..... 209
9.2 Annual Highlights of International Cyberspace Governance ..... 210
9.2.1 The Outbreak of COVID-19 Exacerbates the Uncertainty and Vulnerability in International Cyberspace Governance ..... 210
9.2.2 Geopolitics Enormously Affects the Direction of International Cyberspace Governance ..... 211
9.2.3 The Model of International Cyberspace Governance Needs to Be Adjusted and Reformed ..... 211
9.3 New Progress Made in Issues Related to International Cyberspace Governance ..... 212
9.3.1 Rules for International Cyberspace Governance Continue to Advance ..... 212
9.3.2 The Game of Making Rules for Digital Economy Intensifies ..... 213
9.3.3 Consensus is Gradually Formed in the Governance of Cyber Content ..... 214
9.3.4 The Governance and Standard-Making of Information Technology Keeps Improving ..... 215
9.3.5 Talks on the Security of ICT Supply Chain Deepen ..... 215
9.3.6 International Community Strives to Bridge the Digital Divide ..... 217
9.4 Cyberspace Governance in Some Typical Countries and Regions ..... 217
9.4.1 The United States ..... 218
9.4.2 China ..... 219
9.4.3 Russia ..... 220
9.4.4 The European Union ..... 220
9.4.5 Japan ..... 221
9.4.6 The United Kingdom ..... 222
9.4.7 France ..... 223
9.4.8 Germany ..... 223
9.4.9 India ..... 224

9.4.10 Australia .......... 224
9.4.11 Latin America .......... 225
9.4.12 Africa .......... 226

**Postscript** .......... 227

# Chapter 1
# Overview

## 1.1 Overall Trends in World Internet Development in 2020

The world today is marked by changes unseen in a century, and Internet development is facing new opportunities and challenges. In 2020, the COVID-19 Pandemic broke out and spread at a global scale, which enormously impacted the global economy and society. Internet played an increasingly important role. Digital economy became an important force that hedged against the impact of the COVID-19 Pandemic, reshaped the economic system, and enhanced governance capacity. All countries in the world vigorously promoted the construction of information infrastructure represented by 5G, artificial intelligence, Internet of Things and so on. A new round of scientific & technological revolution and industrial transformation represented by new-generation information technology accelerated the evolution. Global economy further transformed and upgraded toward digital economy, and the strong rise of digital technology promoted the deep industrial integration. Internet media achieved diversified development, with more diverse and richer cyber content and faster cyber cultural exchange and collision. All countries in the world attached great importance to the construction and protection of cybersecurity, constantly improved their own legal systems in the field of Internet, constantly strengthened Internet governance, and further developed the cybersecurity industry.

Meanwhile, we must notice that there are more uncertainties in the development direction of the world Internet development in the future. As unilateralism and trade protectionism have risen, global technological innovation and exchange & cooperation, industrial chain & supply chain have been restricted and destroyed, and the pace of R&D and construction of information infrastructure has slowed down. During the outbreak of the COVID-19 Pandemic, false information and malicious information on the Internet increased greatly, and data and privacy security worried people. New governance problems derive from the iterative development of information technology, and the militarization of cyberspace becomes more and more evident. This poses more challenges to the international Internet governance. As a consequence, global cyberspace governance system needs to be explored and advanced urgently.



Chinese Academy of Cyberspace Studies, *World Internet Development Report 2020*,
https://doi.org/10.1007/978-981-16-9388-5_1

Facing the profound changes in the international situation and the global impact of the COVID-19 Pandemic, international community should develop, use and govern Internet well, so that Internet can better benefit human. All countries in the world should conform to the trend of the times, undertake the responsibility of development, confront challenges and risks together, jointly promote the global cyberspace governance, and strive to build a community with a shared future in cyberspace.

### *1.1.1 COVID-19 Pandemic Hits Global Economic and Social Development, and Digital Economy is Viewed as a New Engine for the Recovery of Global Economy*

In 2020, changes in global landscape were overlaid with the impact of the COVID-19 Pandemic, and the downward pressure on the world economy continued to increase. World Trade Organization (WTO) predicted that global trade would shrink by 13% to 32% in 2020. International Monetary Fund (IMF) forecasted that the global economy would lessen by 4.4% in 2020.

The outbreak of the COVID-19 Pandemic accelerated the digital transformation of the global economy, and digital economy perhaps became a key force that hedged against the impact of the COVID-19 Pandemic, reshaped the economic system and enhanced governance capacity. In Digital Economy Report 2019 released by United Nations Conference on Trade and Development, it was stated that the scale of digital economy was expected to account for 4.5%-15.5% of world GDP based on its various definitions. As the global practice and development of digital economy suggested, in 2020, various countries continuously advanced the digital industrialization and industrial digitization and promoted the integration and development of digital economy and real economy. The market of basic telecommunications and electronic information manufacturing industry kept growing. Information technology service industry such as big data, artificial intelligence and blockchain developed continuously, and Internet information content service industry grew rapidly. The digitization of manufacturing industry further deepened, the digital transformation and upgrading of service industry accelerated, the digitization of agriculture developed steadily, and Fintech and e-commerce advanced in a deep way. To develop digital economy became an important direction for all countries to develop the economy and promote the economic recovery.

### 1.1.2 Internet Media Achieves Diversified Development, and All Countries Continuously Strengthen Cyber Content Governance

In 2020, the world Internet media developed more diversely, and the number of users remained growing and their activity degree kept rising. New digital media platforms emerged constantly, and streaming media and digital entertainment technologies evolved each day. The market share of digital publishing and network literature continuously expanded, and the use case of information technology in the field of media diversified. New media, new technology and new application enlarged the space for people to communicate and provided them with rich cultural resources; yet, they profoundly altered the ecosystem of cyber public opinions and the rule of information dissemination. This produced enormous challenges to cyber content governance. In particular, in the COVID-19 Pandemic, false information and cyber violence & terrorism emerged endlessly and varied greatly, which seriously affected the social order and daily life. António Guterres, Secretary-General of the United Nations, stated that wrong and false information on the COVID-19 Pandemic constantly spread and directly impacted people's production and life in the world. In the recent research report, UNESCO defined the phenomenon that false information intensified the spread of pandemic as "infodemic".[1]

Strengthening cyber content governance has become a global consensus. More and more countries start to use legislative and technological means to heighten cyber content regulation of social platforms and maintain good cyber ecology. For example, the United Kingdom released Online Harm White Paper and proposed to enhance the regulation of cyber platforms such as social media legislatively. Canada issued Digital Charter to strengthen the regulation of false news and hatred speeches. Singapore passed Protection from Online Falsehoods and Manipulation Act 2019, which aimed to penalize platforms and individuals that maliciously disseminated information and attempted to harm public interests.

### 1.1.3 All Countries in the World Actively Deploy New-Type Infrastructure and Accelerate the Construction of Information Infrastructure

In 2020, the construction of new-type infrastructure represented by 5G, artificial intelligence, Internet of Things and industrial Internet accelerated. 5G entered a new stage of fully commercial uses, and all countries actively deployed 5G network. The 5G terminal ecosystem constantly diversified, the scale of 5G users rapidly expanded, and the distribution of 5G spectrum was actualized and refined. Global data centers grew steadily, the competition pattern on cloud computing intensified, and

---

[1] https://en.unesco.org/sites/default/files/disinfodemic_deciphering_covid19_disinformation.pdf.

blockchain quickened continuously. All countries continued to strengthen the layout of AI infrastructure and increase investment into AI platform, which ushered in a stage of rapid development for AI construction. Globally, the commercial deployment of IPv6 was promoted, with sustainable increase of IPv6 assigned number. Global IoT platform market remained active, and all countries positively deployed the digital construction in the industrial field.

Meanwhile, all countries in the world have started to invest into the R&D of next-generation information infrastructure. Countries represented by the United States, China, Japan, South Korea and Finland have successively deployed the R&D of 6G technology. In addition, all countries increased investment into fields such as Satellite Internet, quantum network and wider IoT gradually and took them as key strategic development direction, so as to quicken the transformation and upgrading of economic and social fields.

### 1.1.4 Innovation in the Emerging Internet Technology Shows Vitality, and Various Countries Strive to Bridge the Digital Divide

At present, a new round of scientific & technological revolution and industrial transformation speeds up. New technologies, new applications and new business forms in Internet keep growing. Global major economies and leading corporations continue to deploy and increase investment in order to promote the further development of the emerging Internet technology. According to 2019 EU Industrial R&D Scoreboard released by EU, the top 2,500 corporations in the world continuously increase investment into Internet technology, concentrating the R&D direction on the fields of ICT and healthcare. Global innovation in the Internet technology shows vitality and constantly promotes the rapid economic and social development.

Global development encounters big challenges. Under such circumstance, some governments and corporations have strengthened cooperation and worked to bridge the digital divide. As the survey data of International Telecommunication Union (ITU) suggests, there are now 4.1 billion people worldwide who obtain access to Internet services, yet there are about 3.6 billion people "offline" to the Internet, most of which live in the least developed countries. The penetration rate of Internet reaches as high as 87% in developed countries but as the lowest as 28.2% in Africa.[2]

China actively advocates the Digital Silk Road and plays an important role in bridging the global digital divide. By June 2019, China has signed the memorandum of understanding (MOU) with 16 countries on strengthening the construction and cooperation of the Digital Silk Road, memorandum of understanding with 19 countries on bilateral e-commerce cooperation. In May 2020, China Mobile and Facebook announced with telecommunications companies from South Africa, France, Saudi

[2] ITU: "Globally there are 52% of women inaccessible to Internet with enlarged gender digital divide". See https://news.un.org/zh/story/2019/11/1044991.

Arabia and Egypt that they would collaborate on laying 2Africa submarine cable to provide services to the African continent and the Middle East. The project is the submarine cable that covers the African continent most extensively. Alipay also establishes direct business or runs business via local payment platforms in more than 40 European and Asian countries.

### 1.1.5 Cybersecurity Threats Become More Prominent, and All Countries Constantly Strengthen Cybersecurity Protection

Under the background of the fast development of information technology, the problem of cybersecurity becomes more prominent. All countries in the world continue to strengthen cybersecurity protection capacity and increase investment into the field of cybersecurity. However, the development of security technology lags behind the pace of malicious use of digital technology. For instance, the frequency, influence and complexity of cybersecurity threats are constantly upgraded.[3] Cybersecurity problems such as Advanced Persistent Threat (APT) attacks, malware, data leakage and ransomware occur frequently in the world, and cybersecurity threats gradually display a multi-level, multi-dimensional and cross-domain superposition trend.

All countries in the world take cybersecurity as an important national strategy and issue cybersecurity-related strategic plans successively. For example, the United States formulated National Security and Personal Data Protection Act of 2019 (Draft), Poland issued Poland Cybersecurity Strategy (2019–2024), Brazil released National Cybersecurity Strategy (2020–2023), Singapore issued Master Plan for Operational Technology on Cybersecurity, Ireland published National Cybersecurity Strategy 2019–2024, and so on. However, as cybersecurity plays a rising strategic role in national security and the trend of militarization in cyberspace becomes more fierce, asymmetric threats, unilateral transparency threats and hegemonism threats that most countries face constantly intensify, and global rules on cybersecurity recognized by all countries need to be formulated.

Under such circumstance, in September 2020, at the international symposium themed on "Seizing Digital Opportunities and Seeking Cooperation and Development", China put forward Global Initiative on Data Security and proposed to protect data security. It underscored that all parties should strengthen communication, deepen dialogue and cooperation on the basis of mutual respect, and join hands to forge a community with a shared future in cyberspace featuring peace, security, openness, cooperation and order.

---

[3] World Economic Forum: Why we must treat cybersecurity like public good, August 2019. See https://www.weforum.org/agenda/2019/08/we-must-treat-cybersecurity-like-public-good/.

### 1.1.6 Internet Facilitates the Exchange and Cooperation Among Countries in the World, and International Cyberspace Governance Needs to Be Explored and Promoted

In face of a series of new problems and challenges caused by Internet development, it becomes a broad consensus in the international community to strengthen international cooperation on cyberspace and jointly handle the risks and challenges on the Internet. All countries in the world continue to strengthen communication, deepen dialogue and cooperation, improve the level and capacity of cyberspace governance, and join hands to forge a community with a shared future in cyberspace featuring peace, security, openness, cooperation and order. Cyber sovereignty is widely recognized by more and more countries. All countries actively explore rules and models on Internet development and governance that are inclusive to multilateral development and play a more important part in protecting cyberspace order and security, which advances international Internet governance toward a fairer and more reasonable direction.

In the COVID-19 Pandemic, Internet played an important role in enhancing exchange and cooperation among various countries and promoting their economic development. In March 2020, United Nations Security Council held its first video conference in history to examine and consider related issues. In May 2020, the 73rd World Health Assembly was held in the form of web-based tele-conference, calling on to strengthen international cooperation to combat the COVID-19 Pandemic. In September 2020, China International Fair for Trade in Services was held. Themed on "Global Services, Mutual Benefit and Sharing", it stressed that all countries should conform to the trend of digital, networked and intelligent development and work together to eliminate the digital divide.

However, it is noticeable that the deep adjustment and change of international landscape and world order, together with the outbreak of the COVID-19 Pandemic globally, exacerbate the uncertainty and vulnerability of international cyberspace governance. Perhaps the COVID-19 Pandemic will lead to the less concern and fewer resources that various countries invest into international cyberspace governance. Geopolitics will profoundly impact the trend of international cyberspace governance, the model of which will face adjustment and change. Therefore, all countries should conform to the trend of the times, undertake their responsibility of development, jointly handle risks and challenges, advance International cyberspace governance together, and strive to promote the construction of a community with a shared future in cyberspace to achieve the goals of equality and respect, innovative development, openness and sharing, and security and order in cyberspace.

## 1.2 Internet Development in Representative Countries

In 2017, World Internet Development Report established Global Internet Development Index (GIDI) System. In 2020, GIDI System selects and analyzes 48 representative countries in Five Continents to present the latest Internet development in the continents and the world. The 48 representative countries are listed as below:

America: The United States, Canada, Brazil, Argentina, Mexico, Chile, Cuba.

Asia: China, Japan, South Korea, Indonesia, India, Saudi Arabia, Turkey, The United Arab Emirates, Malaysia, Singapore, Thailand, Israel, Kazakhstan, Vietnam, Pakistan, Iran.

Europe: The United Kingdom, France, Germany, Italy, Russia, Estonia, Finland, Norway, Spain, Switzerland, Denmark, Netherlands, Portugal, Sweden, Ukraine, Poland, Ireland, Belgium.

Oceania: Australia, New Zealand.

Africa: South Africa, Egypt, Kenya, Nigeria, Ethiopia.

### *1.2.1 Index Construction*

Global Internet Development Index (GIDI) comprehensively measures and reflects a country's Internet development from six aspects: infrastructure, innovation capacity, industrial development, Internet application, cybersecurity, and cyberspace governance. In the past two years, indicator system consisted of 6 first-level indicators, 12–15 s-level indicators and many third-level indicators. The past two-year research suggested that metadata of various indicators were available. Considering that, Global Internet Development Index (GIDI) in 2020 maintains 6 first-level indicators, and revises second-level indicators (17) and third-level indicators (34).

### *1.2.2 GIDI System*

Infrastructure, innovation capacity, industrial development, Internet application, cybersecurity, and cyberspace governance are main factors that affect Internet development. The weights of these factors are basically consistent with 2019, and data sources used are modified slightly. See Table 1.1 for Global Internet Development Index (GIDI) System.

**Table 1.1** Global Internet development index (GIDI) system

| First-level indicators | Second-level indicators | Third-level indicators | Indicator description | Data sources |
|---|---|---|---|---|
| 1. Infrastructure | 1.1 Fixed infrastructure | 1.1.1 Average download rate of fixed broadband network | Reflecting the average download rate of fixed broadband users over a given period of time in different countries | Statistical data of global digital reports (Global Web Index, etc.), 2019 |
| | | 1.1.2 IPv6 | Reflecting the deployment of IPv6 | Global IPv6 Supporting Degree White Paper 2020 |
| | 1.2 Mobile infrastructure | 1.2.1 Average download rate of mobile broadband network | Reflecting the average download rate of mobile broadband users over a given period of time in different countries | Statistical data of global digital reports (Global Web Index, etc.), 2019 |
| | | 1.2.2 Infrastructure of mobile network | Reflecting the infrastructure construction of mobile network in different countries | Statistical data of global digital reports (Global Web Index, etc.), 2019 |
| | | 1.2.3 Charge burden of mobile network | Reflecting the ratio of mobile network charge in GNI | Database of International Telecommunication Union, 2019 |
| | 1.3 Application infrastructure | 1.3.1 Number of super computer | Reflecting the number of super computer in different countries | Top 500, 2019 |
| 2. Innovation capacity | 2.1 ICT patent application | 2.1.1 Number of ICT patent application | Reflecting the level and capacity of ICT patent application in different countries | Database of Organization for Economic Co-operation and Development, 2017 |
| | 2.2 Innovation and development capacity | 2.2.1 Ratio of R&D investment in GDP | Reflecting the ratio of R&D investment in GDP in different countries | The World Bank, 2016 |

(continued)

**Table 1.1** (continued)

| First-level indicators | Second-level indicators | Third-level indicators | Indicator description | Data sources |
|---|---|---|---|---|
| | | 2.2.2 Innovation driving force | Reflecting the innovation in diverse development, R&D and marketization of innovation achievement in different countries | World Economic Forum, 2019 |
| | 2.3 Innovation potential | 2.3.1 Ratio of talents with digital skills | Reflecting the ratio of talents with digital skills in the total population in different countries | World Economic Forum, 2019 |
| 3. Industrial development | 3.1 Industrial development environment | 3.1.1 Intellectual property rights protection | Reflecting intellectual property rights protection in different countries | Statistical data of World Economic Forum, 2018 |
| | | 3.1.2 Capacity of participating in globalization | Reflecting the level of participating in globalization in terms of economic, social and political dimensions in different countries | Swiss Economic Institute, 2019 |
| | 3.2 Digital industry | 3.2.1 Ratio of ICT value added | Reflecting the ratio of ICT industrial value added in GDP in different countries | United Nations Conference on Trade and Development, 2017 |
| | | 3.2.2 ICT service export ratio | Reflecting the ratio of ICT service export scale in that of domestic services in different countries | Statistical data from World Development Indicators (WDI) of The World Bank, 2018 |
| | | 3.2.3 ICT product export ratio | Reflecting the ratio of ICT product export scale in that of domestic products in different countries | Statistical data from World Development Indicators (WDI) of The World Bank, 2018 |

(continued)

**Table 1.1** (continued)

| First-level indicators | Second-level indicators | Third-level indicators | Indicator description | Data sources |
|---|---|---|---|---|
| | | 3.2.4 Number of unicorns in digital industry | Reflecting the number of corporations with the market value over $1 billion in digital industry in different countries | Statistical data of CB Insights, 2020 |
| | | 3.2.5 Creation of mobile applications | Reflecting the creation of mobile applications in different countries | Statistical data of International Monetary Fund, 2019 |
| | 3.3 Economic effect of digital industry | 3.3.1 Degree of enterprise's digital transformation | Reflecting the use of ICT to improve commercial model in different countries | Statistical data of World Economic Forum, 2018 |
| | | 3.3.2 Impact of digital economy on new organizational model | Reflecting the use of ICT to improve organizational model, such as building virtual team and telework, in different countries | Statistical data of World Economic Forum, 2018 |
| 4. Internet application | 4.1 Person-level applications | 4.1.1 Number of Internet users | Reflecting the total number of Internet users in different countries | Statistical data of global digital reports (by Global Web Index, etc.), 2019 |
| | | 4.1.2 Online time at social media | Reflecting online time at social media in different countries | Statistical data of global digital reports (Global Web Index, etc.), 2019 |
| | | 4.1.3 Online activity degree | Assessing the numbers of international top-level domains, national & regional top-level domains and Wikipedia editors | Statistical data of ZookNIC and the United Nations, 2019 |

(continued)

**Table 1.1** (continued)

| First-level indicators | Second-level indicators | Third-level indicators | Indicator description | Data sources |
|---|---|---|---|---|
| | 4.2 Business-level applications | 4.2.1 ICT used in B2B transaction | Reflecting enterprise's capacity in using ICT in B2B transaction in different countries | Statistical data of World Economic Forum, 2018 |
| | | 4.2.2 ICT used in B2C transaction | Reflecting the role of Internet in e-commerce | Statistical data of World Economic Forum, 2018 |
| | 4.3 Government-level applications | 4.3.1 Online service index | Reflecting the quality of online service provided by government websites in different countries | Statistical data of the United Nations, 2019 |
| | | 4.3.2 E-participation index | Reflecting the online communication between citizens and government in different countries | Statistical data of the United Nations, 2019 |
| | | 4.3.3 Government data openness index | Reflecting the level of government data openness in different countries | Statistical data of the United Nations, 2019 |
| 5. Cybersecurity | 5.1 Cybersecurity facilities | 5.1.1 Number of secure network servers per 1 million people | Reflecting the number of secure network servers per 1 million people in different countries | Statistical data of database of The World Bank, 2018 |
| | 5.2 Cybersecurity industries | 5.2.1 Number of Cybersecurity Enterprises Top 150 | Reflecting the number of Hot Cybersecurity Enterprises Top 150 in different countries | List of Hot Cybersecurity Enterprises 150 published by Cybersecurity Ventures, 2019–2020 |

(continued)

**Table 1.1** (continued)

| First-level indicators | Second-level indicators | Third-level indicators | Indicator description | Data sources |
|---|---|---|---|---|
| | 5.3 Level of cybersecurity | 5.3.1 Times of cyberattacks | Reflecting mobile devices infected with malware, financial malicious attacks, computers infected with malware, Telnet attacks, the level of encrypted mining attacks and the state of cyber defense, the latest legislation on cybersecurity and so on in different countries | Global Cybersecurity Survey Report by Comparitech, 2019 |
| 6. Cyberspace governance | 6.1 Internet governance | 6.1.1 Related organizations in Internet governance | Reflecting related organizations in Internet governance in different countries, in such specific affairs as policy, security, protection of critical information infrastructure, Computer Emergency Response Team (CERT), criminal, and protection of consumer | Referring to achievement of foreign researches and inviting experts and scholars in relevant fields to conduct comprehensive evaluation |
| | | 6.1.2 Policies and regulations in Internet governance | Reflecting Internet affairs or the making of policies and regulations related to Internet Service Provider (ISP) in different countries | Referring to achievement of foreign researches and inviting experts and scholars in relevant fields to conduct comprehensive evaluation |

(continued)

**Table 1.1** (continued)

| First-level indicators | Second-level indicators | Third-level indicators | Indicator description | Data sources |
|---|---|---|---|---|
| | 6.2 Participation in international governance | 6.2.1 Participation in international meetings in Internet governance | Reflecting participation in international conferences on cyberspace in different countries, including bilateral meetings, multilateral meetings and other forums | Referring to achievements of foreign researches and inviting experts and scholars in relevant fields to conduct comprehensive evaluation |
| | | 6.2.2 Leading or participating in cyber capacity construction | Reflecting helping others in cyber capacity construction, including technological assistance, policy guidance or project training in different countries | Referring to achievements of foreign researches and inviting experts and scholars in relevant fields to conduct comprehensive evaluation |

## 1.2.3 *Analysis of Results*

Through the calculation of indicators, the scores of Internet Development Index in the 48 countries are obtained, as shown in Table 1.2. As it signals, the United States and China take the lead over other countries in Internet development, European countries display powerful strength and balanced development in the Internet, and Latin America and Sub-Saharan Africa progress remarkably in Internet development.

### 1.2.3.1 All Countries Actively Build and Deploy Information Infrastructure, and 5G Scaled Deployment and Commercial Uses Accelerate

All countries further improve fixed infrastructure and mobile infrastructure, with the average download rates of fixed and mobile networks gradually increased. Singapore takes the lead in fixed infrastructure and mobile infrastructure, with higher download rate, wider network coverage and lower network charge. Global IPv6 deployment advances rapidly. Developed countries in North America and Europe maintain high level and stable development. Some developing countries, such as India, Malaysia and Vietnam, develop rapidly and show strong momentum. The United States, China and Japan possess distinctive advantages in global development of super computer. The United States, Canada and other countries play a leading role in the application capacity and software level of super computer, while China boasts in the construction quantity and fund investment of super computer.[4]

In 2020, many countries have realized commercial uses of 5G, and 5G construction has been further accelerated and expanded. Influenced by the COVID-19 Pandemic, many countries delay the issuance of 5G licenses, yet they do not slow the coming of 5G. As of May 2020, China has opened more than 200,000 5G base stations, which has increased by 15,000 a week. 5G network of Deutsche Telekom has already covered more than 1,000 cities and towns and served a population of 40 million. In Netherlands, 5G network has already covered half of the country. KDDI and Softbank, Japanese telecom operators, have actively promoted infrastructure sharing and accelerated the construction of 5G networks in rural areas.

The 48 countries further improve the construction of information infrastructure; however, there is a gap in network universalization rate among them. According to the statistical data of Digital in 2020, the number of global Internet users continues to grow, reaching around 4.54 billion and realizing a universalization rate of 59%, up by nearly 300 million over 2019. Specifically, Nordic area boasts the highest Internet universalization rate, reaching as high as 95%. Internet universalization rate in Europe is higher than other regions in the world, with 92% in Western Europe, 83% in Southern Europe and 78% in Eastern Europe. Internet universalization rate in Asia commonly maintains at around 60%, where East Asia takes the lead and South Asia lags behind. In African continent where Internet infrastructure is relatively weak,

---

[4] https://www.36kr.com/p/1724984672257.

**Table 1.2** Scores of Internet development index in 48 countries

| Rank | Country | Score |
|---|---|---|
| 1 | The United States | 66.14 |
| 2 | China | 55.17 |
| 3 | Germany | 52.47 |
| 4 | The United Kingdom | 52.35 |
| 5 | Singapore | 52.22 |
| 6 | Sweden | 52.02 |
| 7 | Switzerland | 51.77 |
| 8 | France | 51.73 |
| 9 | Canada | 51.29 |
| 10 | Israel | 51.28 |
| 11 | South Korea | 51.25 |
| 12 | The Netherlands | 51.01 |
| 13 | Japan | 50.94 |
| 14 | Denmark | 50.39 |
| 15 | Finland | 49.95 |
| 16 | Australia | 48.34 |
| 17 | Belgium | 48.33 |
| 18 | Spain | 48.03 |
| 19 | Estonia | 48.03 |
| 20 | Ireland | 47.88 |
| 21 | Norway | 47.60 |
| 22 | New Zealand | 47.38 |
| 23 | Italy | 46.62 |
| 24 | India | 46.41 |
| 25 | Portugal | 46.41 |
| 26 | Russia | 46.34 |
| 27 | Malaysia | 46.05 |
| 28 | The United Arab Emirates | 45.69 |
| 29 | Poland | 45.24 |
| 30 | Thailand | 44.40 |
| 31 | Saudi Arabia | 43.94 |
| 32 | Brazil | 43.82 |
| 33 | Vietnam | 43.71 |
| 34 | Turkey | 43.04 |
| 35 | Mexico | 43.03 |
| 36 | Chile | 42.81 |
| 37 | Indonesia | 42.78 |

(continued)

**Table 1.2** (continued)

| Rank | Country | Score |
|---|---|---|
| 38 | Ukraine | 42.75 |
| 39 | South Africa | 42.51 |
| 40 | Argentina | 42.04 |
| 41 | Egypt | 39.89 |
| 42 | Iran | 39.81 |
| 43 | Kenya | 39.53 |
| 44 | Kazakhstan | 39.47 |
| 45 | Pakistan | 39.03 |
| 46 | Nigeria | 35.82 |
| 47 | Ethiopia | 34.11 |
| 48 | Cuba | 31.00 |

Internet universalization rate has great potential, 60% in South Africa (developing the best) and 22% in Central Africa (the lowest). See Table 1.3 for Scores of Information Infrastructure in 48 Countries.

#### 1.2.3.2 Global Scientific and Technological Innovation Keeps Growing, and Global Innovation Landscape Presents Multi-Polarization Development

Digital boom leads the trend of global scientific & technological innovation. In "10 Breakthrough Technologies in 2020" List released by MIT Technology Review, emerging network information technologies take up a large part, including anti-hacker Internet, quantum supremacy, digital currency, satellite mega-constellations, Micro AI and differential privacy.

All countries actively promote innovation and development and increase investment into technological innovation year by year. According to The Global Competitiveness Report 2019, the leading countries in innovation capacity, such as the United States, Germany, Japan, South Korea and Sweden, invest a great deal into R&D, and emerging technologies like artificial intelligence, quantum computing, 5G and blockchain become the key field in their governmental investment. In terms of corporate R&D investment, as 2019 EU Industrial R&D Scoreboard suggests, high-tech fields such as ICT production & service, health and automobile rank the top in corporate R&D investment.

Global scientific & technological innovation landscape presents a trend of multi-polarization. According to Global Innovation Index 2019, Switzerland, Sweden, the United States and other countries maintained their leading position, and advanced economies, or other European countries (like the Netherlands and Germany) and Asian countries (like Singapore and South Korea), lie in top ranking. Some middle-income economies come to rise rapidly. China, the United Arab Emirates, Vietnam,

**Table 1.3** Scores of information infrastructure in 48 countries

| Rank | Country | Score |
|---|---|---|
| 1 | Singapore | 6.00 |
| 2 | China | 5.71 |
| 3 | The United States | 5.54 |
| 4 | South Korea | 5.47 |
| 5 | Switzerland | 5.39 |
| 6 | Canada | 5.13 |
| 7 | France | 5.10 |
| 8 | Norway | 5.02 |
| 9 | The Netherlands | 4.98 |
| 10 | Sweden | 4.80 |
| 11 | The United Arab Emirates | 4.78 |
| 12 | New Zealand | 4.66 |
| 13 | Denmark | 4.63 |
| 14 | Belgium | 4.61 |
| 15 | Japan | 4.60 |
| 16 | Thailand | 4.56 |
| 17 | Portugal | 4.50 |
| 18 | Finland | 4.44 |
| 19 | Germany | 4.37 |
| 20 | Spain | 4.36 |
| 21 | The United Kingdom | 4.01 |
| 22 | Malaysia | 4.00 |
| 22 | Australia | 4.00 |
| 24 | Estonia | 3.99 |
| 25 | Ireland | 3.92 |
| 26 | Poland | 3.89 |
| 27 | Saudi Arabia | 3.86 |
| 28 | Israel | 3.78 |
| 29 | Chile | 3.71 |
| 30 | Italy | 3.64 |
| 31 | Brazil | 3.57 |
| 32 | Vietnam | 3.50 |
| 33 | Mexico | 3.40 |
| 34 | Russia | 3.28 |
| 35 | India | 3.24 |
| 36 | Argentina | 3.21 |
| 37 | Turkey | 3.06 |

(continued)

**Table 1.3** (continued)

| Rank | Country | Score |
|---|---|---|
| 38 | Ukraine | 2.95 |
| 39 | Kazakhstan | 2.89 |
| 40 | South Africa | 2.85 |
| 41 | Kenya | 2.67 |
| 42 | Egypt | 2.64 |
| 43 | Iran | 2.55 |
| 44 | Indonesia | 2.53 |
| 45 | Pakistan | 2.28 |
| 46 | Nigeria | 2.16 |
| 47 | Ethiopia | 2.05 |
| 48 | Cuba | 1.60 |

Thailand, India and Iran display great potential for development. Among them, China ranks top on the score of third-level indicators like the number of ICT patent application and ratio of talents with digital skills. India is the most innovative and dynamic economy in Central Asia and South Asia. It not only possesses advantages in ICT service export and ratio of talents with digital skills, but also owns the world-leading science & technology clusters in Bangalore, Mumbai and New Delhi. See Table 1.4 for Scores of Internet Innovation Capacity in 48 Countries.

#### 1.2.3.3 Digital Industry in Eurasia Grows Strongly, and the Pace of the Digital Transformation of Manufacturing Industry Accelerates

Digital industry serves as a powerful support for the development of digital economy. As emerging digital markets develop in China and India, the strong growth trend of digital industry commences to shift to Eurasia. In the export of ICT services, India, China, Argentina, Finland, Israel, Sweden and Ireland account for a relatively large share. In ICT product exports, Mexico, China, Malaysia, Singapore and Vietnam account for a relatively large share. In the number of unicorns in digital industry, the United States and China take the lead, and the United Kingdom and India follow them.

All countries compete in and promote the development of digital industry, and the digital transformation of manufacturing industry becomes a hot field. Countries and regions such as the United States, South Korea, Germany, China and EU all have launched strategies and policies on the industrial development to seize the dominant role in future development. With the continuous advance of the digital transformation in manufacturing industry, manufacturing enterprises enhance their digital capacities remarkably and increase their operation efficiency and product-innovation speed

**Table 1.4** Scores of Internet innovation capacity in 48 countries

| Rank | Country | Score |
|---|---|---|
| 1 | The United States | 9.50 |
| 2 | Germany | 9.24 |
| 3 | Japan | 9.15 |
| 4 | South Korea | 9.09 |
| 5 | Sweden | 8.88 |
| 6 | China | 8.75 |
| 7 | The United Kingdom | 8.72 |
| 8 | Switzerland | 8.66 |
| 8 | France | 8.66 |
| 10 | Israel | 8.57 |
| 11 | The Netherlands | 8.52 |
| 12 | Canada | 8.48 |
| 13 | Singapore | 8.32 |
| 14 | Denmark | 8.26 |
| 15 | Australia | 8.15 |
| 16 | Belgium | 8.09 |
| 17 | Italy | 7.88 |
| 18 | Norway | 7.82 |
| 19 | Ireland | 7.82 |
| 20 | Spain | 7.76 |
| 21 | Finland | 7.54 |
| 22 | India | 7.45 |
| 23 | Russia | 7.40 |
| 24 | Malaysia | 7.32 |
| 25 | New Zealand | 7.30 |
| 26 | Portugal | 7.03 |
| 27 | Poland | 6.99 |
| 28 | Saudi Arabia | 6.98 |
| 29 | The United Arab Emirates | 6.87 |
| 30 | Turkey | 6.83 |
| 31 | Estonia | 6.74 |
| 32 | South Africa | 6.63 |
| 33 | Mexico | 6.59 |
| 34 | Ukraine | 6.44 |
| 35 | Thailand | 6.29 |
| 36 | Chile | 6.24 |
| 37 | Iran | 6.04 |

(continued)

**Table 1.4** (continued)

| Rank | Country | Score |
|---|---|---|
| 38 | Brazil | 6.03 |
| 39 | Argentina | 6.00 |
| 40 | Kenya | 5.84 |
| 41 | Indonesia | 5.82 |
| 42 | Pakistan | 5.78 |
| 43 | Egypt | 5.75 |
| 44 | Vietnam | 5.58 |
| 44 | Nigeria | 5.58 |
| 46 | Kazakhstan | 5.44 |
| 47 | Ethiopia | 5.26 |
| 48 | Cuba | 3.90 |

continuously. See Table 1.5 for Scores of Internet Industrial Development in 48 Countries.

#### 1.2.3.4 Person-Level Internet Applications Become More Active, and Business-Level and Government-Level Internet Applications Gradually Improve

Stay-at-home economy caused by the COVID-19 Pandemic further promotes the level of person-level Internet applications. New business models and forms such as online medical care, online education, digital entertainment and digital life thrive. According to Global Digital Reports co-released by We Are Social and Hootsuite, by June 2020, there were 3.96 billion social media users in the world, accounting for 51% of the total population and increasing by 10% year on year.

The digital transformation of traditional enterprises calls for the support of software. At present, North America and Europe mainly dominate the business-level software market, and their enterprises display a relatively high digital level. The United States owns a group of famous industrial software corporations in the world, such as Oracle, Microsoft, General Electric, IBM and Salesforce, which dominate and control the high-end links in the industrial chain of global industrial software, including 90% of the world's operating systems, database management software and most universal suite software and high-end industrial software. In Europe, developed countries like Germany and France rely on deep industrial accumulation and play a dominant role in R&D design, production control, business management and other sub-fields. They own many famous industrial software corporations in the world, like Siemens, SAP, Dassault and ABB. In Asia, developed countries like Japan and South Korea possess strong advantages in the field of industrial software. They master some core technologies and standards in the field of global industrial software through the common development with the world's advanced industrial system. India and China

**Table 1.5** Scores of Internet industrial development in 48 countries

| Rank | Country | Score |
|---|---|---|
| 1 | The United States | 18.00 |
| 2 | China | 15.14 |
| 3 | Israel | 14.12 |
| 4 | Finland | 13.88 |
| 5 | Singapore | 13.39 |
| 6 | Sweden | 13.27 |
| 7 | Estonia | 12.98 |
| 8 | The United Kingdom | 12.82 |
| 9 | Ireland | 12.79 |
| 10 | Switzerland | 12.55 |
| 11 | The Netherlands | 12.45 |
| 12 | Germany | 12.41 |
| 13 | Denmark | 12.36 |
| 14 | France | 12.31 |
| 15 | South Korea | 12.23 |
| 16 | Canada | 11.92 |
| 17 | Japan | 11.90 |
| 18 | Malaysia | 11.89 |
| 19 | India | 11.65 |
| 19 | Belgium | 11.65 |
| 21 | Vietnam | 11.64 |
| 22 | Spain | 11.53 |
| 22 | Australia | 11.36 |
| 24 | New Zealand | 11.32 |
| 24 | Portugal | 11.32 |
| 26 | The United Arab Emirates | 11.06 |
| 27 | Ukraine | 10.97 |
| 28 | Poland | 10.96 |
| 29 | Italy | 10.93 |
| 30 | Mexico | 10.80 |
| 31 | Chile | 10.79 |
| 32 | Indonesia | 10.73 |
| 33 | Russia | 10.69 |
| 34 | Thailand | 10.68 |
| 35 | Saudi Arabia | 10.51 |
| 36 | Turkey | 10.47 |
| 37 | Brazil | 10.46 |

(continued)

**Table 1.5** (continued)

| Rank | Country | Score |
|---|---|---|
| 38 | Kenya | 10.35 |
| 38 | South Africa | 10.35 |
| 40 | Argentina | 10.30 |
| 41 | Egypt | 10.29 |
| 42 | Norway | 10.28 |
| 43 | Pakistan | 10.20 |
| 44 | Iran | 9.86 |
| 45 | Kazakhstan | 9.83 |
| 46 | Ethiopia | 8.86 |
| 47 | Cuba | 7.70 |
| 48 | Nigeria | 7.50 |

have achieved faster growth in industrial software markets and jointly promoted the market growth with developed countries such as Japan and South Korea.

Government-level Internet applications make remarkable progress, and e-government continues to improve in development level. The United Nations E-Government Survey 2020 suggests that there is often a positive correlation between the development level of e-government and the income level in a nation, yet financial resource is not the only key factor. In terms of continents, European countries remain the leading role in the level of e-government, and 95% countries provide at least 10 types of services. Asian countries show a large gap in e-government development. African countries accelerate the development with a positive sign. See Table 1.6 for Scores of Internet Application in 48 Countries.

#### 1.2.3.5 Investment into Cybersecurity Gains Strong Momentum, and North America, Western Europe and the Asia–Pacific Region Take the Lead in Cybersecurity Level

In the COVID-19 Pandemic, as digital economy achieves growth against the adverse trend, all governments and enterprises in the world attach more attention to cybersecurity. Internet data center (IDC) predicts that in 2020, the total global investment into hardware, software and service market related to cybersecurity will reach $120.28 billion, a year-on-year increase of 10.1%[5] over 2019.

Countries in North America, Western Europe and the Asia–Pacific Region present high level in cybersecurity, among which the United States and Israel remain the leading position over 2019. Global Cybersecurity Index (GCI) published by International Telecommunication Union (ITU) in 2019 compared major countries in terms of legislative measures, technological mechanisms, organizational structures, capacity

[5] https://www.idc.com/getdoc.jsp?containerId=prCHC46140120.

**Table 1.6** Scores of Internet application in 48 countries

| Rank | Country | Score |
|---|---|---|
| 1 | The United States | 14.00 |
| 2 | The United Kingdom | 13.60 |
| 3 | Germany | 13.43 |
| 4 | China | 13.29 |
| 4 | The Netherlands | 13.29 |
| 6 | France | 12.84 |
| 7 | Sweden | 12.82 |
| 8 | Canada | 12.79 |
| 9 | Switzerland | 12.73 |
| 10 | India | 12.59 |
| 11 | Denmark | 12.55 |
| 12 | Australia | 12.54 |
| 13 | Japan | 12.47 |
| 14 | Russia | 12.41 |
| 15 | Finland | 12.38 |
| 16 | Brazil | 12.32 |
| 17 | Israel | 12.29 |
| 18 | Indonesia | 12.27 |
| 19 | Spain | 12.20 |
| 20 | Norway | 12.19 |
| 21 | South Korea | 12.18 |
| 22 | Italy | 12.01 |
| 23 | Vietnam | 11.93 |
| 24 | Poland | 11.89 |
| 25 | Mexico | 11.85 |
| 26 | Singapore | 11.84 |
| 27 | Belgium | 11.78 |
| 28 | Thailand | 11.75 |
| 29 | Turkey | 11.70 |
| 30 | New Zealand | 11.68 |
| 31 | Portugal | 11.64 |
| 32 | Malaysia | 11.63 |
| 32 | Ireland | 11.63 |
| 34 | Ukraine | 11.58 |
| 35 | Saudi Arabia | 11.48 |
| 35 | Estonia | 11.48 |
| 37 | Argentina | 11.44 |

(continued)

**Table 1.6** (continued)

| Rank | Country | Score |
|---|---|---|
| 38 | South Africa | 11.43 |
| 39 | Nigeria | 11.30 |
| 40 | Egypt | 11.28 |
| 41 | Chile | 11.23 |
| 42 | Iran | 11.14 |
| 43 | The United Arab Emirates | 11.12 |
| 44 | Pakistan | 11.11 |
| 45 | Kenya | 10.93 |
| 46 | Kazakhstan | 10.68 |
| 47 | Ethiopia | 9.85 |
| 48 | Cuba | 8.00 |

building and cooperation agreements, revealing that in cybersecurity level,, countries such as the United Kingdom, the United States, France, Estonia, Singapore, Spain and Malaysia scored higher, while Iran, Nigeria, South Africa, Vietnam and India scored lower. Comparitech, a technological corporation, evaluated the cybersecurity status in 60 countries from multi dimensions, including mobile devices infected with malware, financial malicious attacks, computers infected with malware, the state of cyber defense and the latest legislation on cybersecurity. It viewed that in cybersecurity level, countries such as Denmark, Sweden, Germany, Ireland, Japan and Canada scored higher, while Iran, Pakistan, Egypt, Brazil, Vietnam and India scored lower. See Table 1.7 for Scores of Cybersecurity Level in 48 Countries.

#### 1.2.3.6 All Countries Strengthen Domestic Internet Governance, and International Community Actively Carries Out Exchange and Cooperation

As a boom in Internet governance arises, many countries concentrate on data, algorithms, competition, content, tax governance and other issues by strengthening legislation and law enforcement. European Union, China, the United States, India and other countries and regions have issued a series of laws and standards on the protection of personal data and privacy. And European Union has imposed several penalties in accordance with General Data Protection Regulation (GDPR). The United States, Russia, EU member states, India and other countries have regulated and controlled such digital platforms as Google, Facebook and Amazon by means of investigation and fine. More and more countries, such as the United States, China, the United Kingdom, Canada and Singapore, have strengthened the regulation of contents on social media and released related laws and policies to combat false information. European countries such as France, Italy, Spain, the United Kingdom and Austria have started to impose digital service tax on large technological corporations, and

**Table 1.7** Scores of cybersecurity level in 48 countries

| Rank | Country | Score |
|---|---|---|
| 1 | The United States | 9.50 |
| 2 | Israel | 4.74 |
| 3 | The United Kingdom | 4.16 |
| 4 | Denmark | 4.12 |
| 5 | Canada | 4.06 |
| 6 | The Netherlands | 4.01 |
| 7 | Singapore | 4.00 |
| 8 | Germany | 3.98 |
| 9 | Switzerland | 3.96 |
| 9 | Ireland | 3.96 |
| 11 | New Zealand | 3.95 |
| 12 | Estonia | 3.94 |
| 12 | Finland | 3.94 |
| 14 | Australia | 3.82 |
| 14 | Norway | 3.82 |
| 16 | France | 3.79 |
| 17 | Japan | 3.77 |
| 17 | Sweden | 3.77 |
| 19 | Poland | 3.74 |
| 20 | Belgium | 3.73 |
| 20 | Portugal | 3.73 |
| 22 | Spain | 3.71 |
| 23 | Italy | 3.69 |
| 24 | South Africa | 3.68 |
| 25 | Russia | 3.67 |
| 26 | Chile | 3.65 |
| 27 | Ukraine | 3.60 |
| 28 | Malaysia | 3.57 |
| 29 | Turkey | 3.54 |
| 30 | South Korea | 3.52 |
| 31 | Argentina | 3.46 |
| 32 | Brazil | 3.44 |
| 33 | Vietnam | 3.43 |
| 34 | Kazakhstan | 3.42 |
| 35 | The United Arab Emirates | 3.39 |
| 36 | Indonesia | 3.37 |
| 37 | Thailand | 3.35 |

(continued)

**Table 1.7** (continued)

| Rank | Country | Score |
|---|---|---|
| 38 | Iran | 3.30 |
| 39 | China | 3.25 |
| 40 | India | 3.16 |
| 41 | Mexico | 3.11 |
| 42 | Kenya | 3.09 |
| 43 | Saudi Arabia | 3.05 |
| 44 | Cuba | 3.02 |
| 45 | Nigeria | 2.92 |
| 46 | Pakistan | 2.89 |
| 47 | Egypt | 2.81 |
| 48 | Ethiopia | 2.50 |

Asian and Latin American countries have started to discuss how to tax technological giants. As the score related to domestic Internet governance capability suggests, the United States, China, the United Kingdom, Germany and other countries have established more comprehensive organizations on Internet governance and more complete policies and regulations on Internet governance, with better performance.

Internet governance embodies complexity in itself. It fails to handle the various challenges if only relies on the power of government. Therefore, multiple parties and forces, such as governments, platforms, users and industrial associations, should play their roles in Internet governance. In December 2019, the United Nations Internet Governance Forum was held. More than 3,400 delegates from 161 countries attended and discussed political, social, technological and ethical issues related to the Internet. This indicated that all parties enlarged consensus on issues such as data security, privacy protection and the risks produced by cybersecurity, and that international organizations played a rising role in cyberspace governance. The extensive exchange helped to promote human's consensus on future Internet development and explore the values, principles and rules for the shift from the real world to the digital world. See Table 1.8 for Scores of Cyberspace Governance in 48 Countries.

## 1.3 Further Information on Selected Countries

By comparison, as the scores of Internet Development Index in the 48 countries reveal, developed countries/regions in North America, Europe and Asia maintain the high level of Internet development on average. Developing countries/regions in Latin America and Sub-Saharan Africa develop quickly in the Internet. The Report selects 10 countries (the United States, China, the United Kingdom, Singapore, Sweden, Canada, Japan, Australia, Italy and Egypt) and analyzes Internet development in them.

**Table 1.8** Scores of cyberspace governance in 48 countries

| Rank | Country | Score |
|---|---|---|
| 1 | The United States | 9.60 |
| 2 | China | 9.04 |
| 2 | Japan | 9.04 |
| 2 | The United Kingdom | 9.04 |
| 2 | France | 9.04 |
| 2 | Germany | 9.04 |
| 7 | Canada | 8.90 |
| 7 | Russia | 8.90 |
| 7 | Estonia | 8.90 |
| 10 | South Korea | 8.75 |
| 11 | Singapore | 8.67 |
| 12 | The United Arab Emirates | 8.47 |
| 12 | Italy | 8.47 |
| 12 | Norway | 8.47 |
| 12 | Spain | 8.47 |
| 12 | Switzerland | 8.47 |
| 12 | Australia | 8.47 |
| 12 | New Zealand | 8.47 |
| 12 | Denmark | 8.47 |
| 12 | Sweden | 8.47 |
| 12 | Belgium | 8.47 |
| 22 | India | 8.33 |
| 23 | Portugal | 8.19 |
| 24 | Indonesia | 8.05 |
| 24 | Saudi Arabia | 8.05 |
| 26 | Brazil | 7.99 |
| 27 | Thailand | 7.77 |
| 27 | Finland | 7.77 |
| 27 | Israel | 7.77 |
| 27 | The Netherlands | 7.77 |
| 27 | Poland | 7.77 |
| 27 | Ireland | 7.77 |
| 33 | Argentina | 7.63 |
| 33 | Malaysia | 7.63 |
| 33 | Vietnam | 7.63 |
| 36 | South Africa | 7.57 |
| 37 | Turkey | 7.43 |

(continued)

**Table 1.8** (continued)

| Rank | Country | Score |
|---|---|---|
| 38 | Mexico | 7.29 |
| 39 | Chile | 7.21 |
| 39 | Kazakhstan | 7.21 |
| 39 | Ukraine | 7.21 |
| 42 | Egypt | 7.12 |
| 43 | Iran | 6.92 |
| 44 | Cuba | 6.78 |
| 44 | Pakistan | 6.78 |
| 46 | Kenya | 6.64 |
| 47 | Nigeria | 6.36 |
| 48 | Ethiopia | 5.60 |

### *1.3.1 The United States Maintains the Global Leading Role in Comprehensive Strength of Internet*

As a world Internet power, the United States keeps the leading role in global Internet technological innovation and development. In the ranking list on Global Internet Development Index of the Report, the United States ranks the 1st in the world. In specific, the United States ranks the 3rd in information infrastructure index, the 1st in innovation capacity index, the 1st in industrial development index, the 1st in the Internet application index, the 1st in cybersecurity index and the 1st in cyberspace governance index. See Fig. 1.1 for Internet Development Index of the United States.

Yet, the problem of digital divide in the United States remains unresolved. In particular, broadband price constitutes the biggest structural obstacle, and the ownership rate of broadband shows the lowest in low-income communities. In 2019 Networked Readiness Index (NRI) released by World Information Technology and

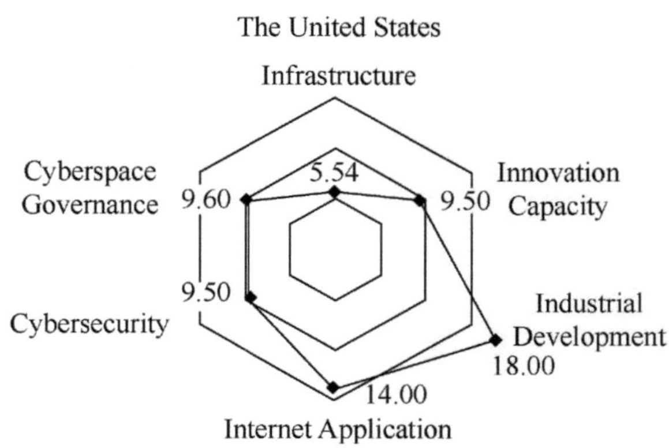


**Fig. 1.1** Internet development index of the United States

Services Alliance (WITSA), the United States scored 80.32, ranking from the 5th (in 2016) to the 8th in the world and underperforming in the Internet access, Internet's impact on society and so on.

The United States ranks the 1st in R&D investment, innovation quality and corporate strength in the world. In 2020 The State of U.S. Science & Engineering published by the United States National Science Foundation, it was revealed that the Unites State continued to occupy the largest share in global R&D investment and R&D intensive industrial output, and awarded the largest number of PhD degrees in science and engineering. According to Global Innovation Index (GII) Report 2019 published by World Intellectual Property Organization (WIPO), the United States ranked the 3rd in innovation capacity and the 1st in innovation quality in the world. There were totally 26 science & technology clusters listed in world top 100 science & technology clusters. In Top 2,000 Global Companies 2020 released by The Forbes in 2020, eight of ten tech companies came from the United States, including Internet giants Apple, Microsoft, Alphabet, Facebook, Intel, IBM, Cisco and Oracle.

The United States strives to dominate in such fields as artificial intelligence and quantum computing. In February 2020, the White House issued American Artificial Intelligence Initiative: Year One Annual Report, emphasizing that it focused on six fields: investing into AI R&D, releasing AI resources, removing barriers to AI innovation, building AI workforce, creating an international environment supportive of U.S. AI innovation, and providing credible AI for governmental services. In addition, in January 2020, the United States Department of Energy launched a plan called "National Quantum Information Science Research Centers" to fund projects on quantum information science. In February 2020, the United States released A Strategic Vision for America's Quantum Networks written by the White House National Quantum Coordination Office, in which it proposed that the United States would develop quantum Internet to ensure quantum information science would benefit the public.

The United States continues to enhance its cybersecurity level and maintain global unilateral advantage. In January 2020, U.S. House Committee on Financial Services passed Cybersecurity and Financial System Resilience Act of 2019, which aimed to ensure that the Federal Reserve System prioritized cybersecurity and modernization. In order to protect the safety of national defense industry, the United States Department of Defense released Cybersecurity Maturity Model Certification (CMMC), which coercively conducted the third-party certification on contractors and replaced the auto-certification model in Defense Federal Acquisition Regulation Supplement. To expand cybersecurity army, the United States Marine Corps is constructing regulatory cyber forces. Despite the unilateral advantage, the United States continues to strengthen the military capacity in cybersecurity. It intensifies the regulation of social media platforms, which arouses global attention. In May 2020, Trump signed an executive order that restricted the "Exemption Clause" in Federal Communications Act for social media. The United States Department of Justice put forward a proposal that urged the Congress to reduce the protection of online platforms and give them greater legal responsibility. These actions in the United States inspired other countries that social media was important resources that concerned political power and security. And all governments started to strengthen their response and formulate relevant policies.

### 1.3.2 *China's Internet Achieves Stable and Rapid Development on the Whole*

In Internet development, China ranks the 2nd and follows the United States in Global Internet Development Index. In specific, China ranks the 2nd in information infrastructure index, the 6th in innovation capacity index, the 2nd in industrial development index, the 4th in the Internet application index, the 39th in cybersecurity index and the 2nd in cyberspace governance index. See Fig. 1.2 for Internet Development Index of China.

China steadily advances Internet infrastructure construction and makes achievements in mobile infrastructure and application infrastructure particularly. Since 2020, China has accelerated the deployment of new-type infrastructure construction. Regions across the country have concentrated on digital infrastructure construction, with many major projects continued to be launched.

(1) The deployment of 5G speeds up. As JINGDATA platform data suggested, by the end of April 2020, there were 2,741 financing cases in 5G field in China, with a total scale of ¥ 250.514 billion and an average single financing of ¥ 91 million. In 2019, three telecom operators, China Mobile, China Unicom and China Telecom, invested more than ¥ 40 billion into 5G, which was expected to reach about ¥ 180 billion in 2020, with an increase of more than 300%.

(2) Data centers are actively expanded. Since March 2020, basic telecom operators, Huawei, Inspur and other server suppliers, as well as Tencent, Baidu, Alibaba and other Internet corporations have increased their deployment successively. They

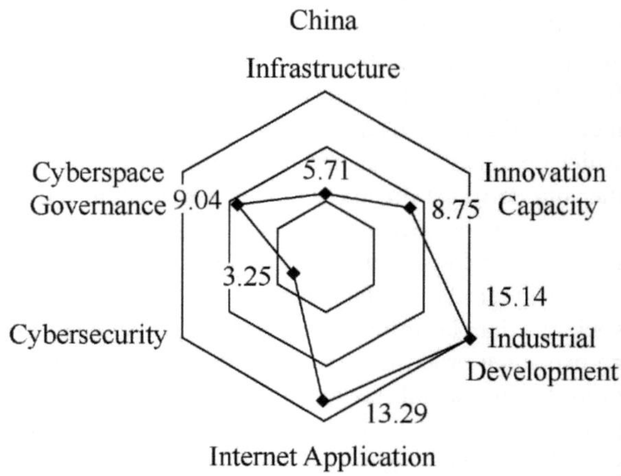


**Fig. 1.2** Internet development index of China

built or expanded data centers in Guangdong-Hong Kong-Macao Greater Bay Area, Zhangye, Lhasa and Chongqing, with an investment of tens of billions of RMB.

(3) Industry-finance cooperation in industrial Internet continues to enlarge. As the data of China Industrial Control Systems Cyber Emergency Response Team suggested, in the first quarter of 2020, there were 40 domestic industrial financing cases in industrial Internet, with the total sum of disclosed financing exceeding ¥ 2 billion and a significant increase over the same period in 2019. In addition, China's Next-Generation Internet IPv6 has been accelerated and upgraded, and network coverage and service capacity have been significantly improved. According to the data of China Internet Network Information Center (CNNIC), by July 2020, China has assigned 1.442 billion IPv6 address users, with 362 million IPv6 active users.

China continuously improves independent innovation capacity. The number of ICT patent application grows year by year, yet China needs to make efforts in R&D investment and the marketization of innovation achievements. According to Global Innovation Index (GII)2019, 18 Chinese science & technology clusters were listed in the world top 100 science & technology clusters. In Bloomberg Innovation Index 2020 released by Bloomberg News (the United States), China performed the most remarkably in patent activities and higher education efficiency, which ranked the 2nd and the 5th respectively. Since 2020, China has made historical breakthroughs in many fields of the integrated circuit industry. As the main force to achieve breakthroughs in advanced process technology in China, Semiconductor Manufacturing International Corporation (SMIC) has completed the mass production of 14 nm process technology. In the first half of 2020, Hisilicon Kirin 710A chip was produced in the form of OEM shipment and the mobile phone Huawei HONOR play 4 T equipped with the chip was opened to the market.

In the COVID-19 Pandemic, China's Internet-related industries rose against the adverse trend, and some industries represented by cloud IT infrastructure developed rapidly. According to the relevant data of Internet data centers (IDC), three of the global top five suppliers in cloud IT infrastructure revenue came from China. In the first quarter of 2020, China accounted for more than 21% of the total global market share, second only to the United States. Information transmission & software and information technology service industries have become important drivers of growth. From January to May 2020, China's information transmission & software and information technology service industries achieved 8.4% growth against the adverse trend. In addition, software industry actively contributed to prevent and control the COVID-19 Pandemic, empowering various industries to resume work and production efficiently. After the outbreak of the COVID-19 Pandemic, many enterprises devoted every effort to ensure the smooth operation of the governmental administration systems at all levels and of the medical information system in major hospitals, so as to guarantee the rapid response to the prevention and control work of the COVID-19 Pandemic. Many enterprises launched free products such as remote communications, cloud-terminal collaboration, online service and remote control to help enterprise's resume production and operation.

China's digital economy has become the main force of stable economic growth and played a leading role in the development of person-level Internet applications. Since

2020, China's mobile Internet has accelerated development, e-commerce, online payment, digital content and online education have grown rapidly, and live broadcast economy has increased in an explosive way. AI use cases are diversified. In use cases such as new retailing, smart medical, pan-entertainment, information content management, intelligent manufacturing and property security, AI has been practiced and verified more fully. The overall level of e-government ranks top in the world. According to The United Nations E-Government Survey 2020, China ranks the 9th from the 29th in the e-participation index, a core indicator that measures the development of national e-government.

In cybersecurity, in May 2019, national standards related to classified protection of cybersecurity system 2.0 were officially released, including Information Security Technology—Baseline for Classified Protection of Cybersecurity, Information Security Technology—Evaluation Requirement for Classified Protection of Cybersecurity, and Information Security Technology—Technical Requirements of Security Design for Classified Protection of Cybersecurity. On October 26 2019, Cryptography Law of the People's Republic of China was adopted by vote, which served as a comprehensive and basic law in the field of Chinese cryptography. Personal Information Protection Law (Draft) will be further amended. In terms of technical standards, Personal Information Security Specification (Revised Edition) has been revised for several rounds and will be completed in the end of 2020. Meanwhile, since 2019, China has successfully launched the campaign that cracks down on App's violation of users' rights and interests. The construction of industrial Internet cybersecurity monitoring and situation awareness platform has been advanced steadily, with coupling effect at national, provincial and corporate levels.

China continues to deepen exchange and cooperation with other countries. In September 2020, China held 2020 China International Fair for Trade in Services and advocated to jointly create an open and inclusive environment for cooperation, to jointly activate the momentum of innovation-led cooperation, and to jointly establish cooperation with mutual benefit and win–win outcome. In the same month, China and Russia released the Joint Statement by the Ministers of Foreign Affairs of the People's Republic of China and the Russian Federation, calling on international community to strengthen cooperation, build consensus, collaborate on handling current threats and challenges, and promote global political stability and economic recovery.

### *1.3.3 The United Kingdom Achieves Balanced Overall Strength in the Internet*

In Global Internet Development Index, the United Kingdom ranks the 4th. In specific, the United Kingdom ranks the 21st in information infrastructure index, the 7th in innovation capacity index, the 8th in industrial development index, the 2nd in the Internet application index, the 3rd in cybersecurity index and the 2nd together in

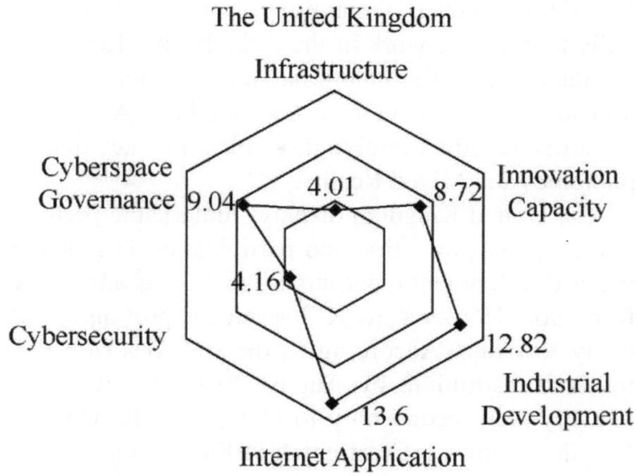


**Fig. 1.3** Internet development index of the United Kingdom

cyberspace governance index. See Fig. 1.3 for Internet Development Index of the United Kingdom.

The United Kingdom strongly supports all-fiber broadband and 5G mobile technology. In recent years, it has made great progress in mobile and fixed network access, yet lagged behind many European countries in all-fiber network coverage. By December 2019, all-fiber broadband coverage rate was only 10% in the United Kingdom.[6] Besides, the low download rates of fixed and mobile broadband networks remain a problem to be solved for the United Kingdom. In January 2020, British Ofcom proposed four plans to accelerate the investment into fiber optic networks, including setting up the upper limit of wholesale prices to encourage competition from new networks, preventing Openreach, a network infrastructure company, from offering substantial discounts to avoid undermining competition, providing greater flexibility for rural areas to encourage the investment, and loosing the regulation of Openreach copper cable networks in areas where all-fiber optic networks were built to help eliminate copper cable network gradually.[7] In 5G, in July 2020, British government announced to launch "5G Create" Fund with 30 million pounds (approximately ¥275 million). As a part of 5GTT (5G Test-platform and Test-plan), the Fund helped to develop innovative projects on 5G technology.[8] According to 2019 Networked Readiness Index, the United Kingdom scored 77.73 in NRI, ranking from the 8th (in 2016) to the 10th. Its disadvantages lied in the Internet application, personal skills, the impact of Internet on economy and life quality and so on.

In the United Kingdom, Internet innovation capacity closely relates to excellent business environment and policy support. It has created a relaxing and flexible regulatory environment for Internet innovation to help domestic technological enterprises attract foreign capital and increase exports. In June 2020, the United

[6] http://www.tlfptw.com/keji/20190926/196.html.

[7] http://paper.cnii.com.cn/article/rmydb_15573_289866.html.

[8] BBC, July 31, 2020.

Kingdom launched Technology & Trade Strategy, stating that it would establish a digital trade network in the Asia–Pacific Region (including Japan, South Korea, Thailand, Australia, Indonesia and Singapore) in the future to help British technological enterprises enter relevant markets. At the same time, the United Kingdom has attracted direct foreign investment into key fields such as 5G, Internet of Things, photonics and Mixed Reality.

The United Kingdom displays remarkable performance and capacity in intellectual property protection and participation in globalization. Digital industry enjoys better development environment and owns advantages in enterprise's digital transformation. However, its ICT service export and product export account for a relatively low rate. According to the statistics of CB Insights, an American market research institution, by August 2020, the number of British unicorns in digital industry was second only to China and the United States, up to 25. The United Kingdom shows great strength in Fintech industry and boasts the capital of global Fintech. In sub-sectors like peer-to-peer lending, online payment, financial data analysis and blockchain, it remains the top level in the world. In World's Top 100 Fintech Companies in 2019 released by KPMG, there were 11 British Fintech companies on the list. Among them, OakNorth, a loan platform, ranked top 10.[9] This benefited from Britain's technological advantages in financial service industry, consumer's strong willingness to try innovative technological products, and the loose regulatory environment British government created.[10]

There are seven British cybersecurity corporations listed in Top 150, second only to the United States and Israel. According to the statistics of British Department for Digit, Culture, Media and Sport (DCMS), by the end of 2019, British cybersecurity industry achieved a total income of about 8.3 billion pounds (approximately 75.4 billion yuan), up by 46% over 2017. Among them, there were more than 1,200 active cybersecurity companies, with around 43,000 employees and total investment of more than 348 million pounds.[11] In June 2020, British Ministry of Defence officially established the 13th Signal Regiment, a force dedicated to cybersecurity, to respond to the growing digital threats from potential adversaries and promote cybersecurity as one of the pillars in national defense strategy. The 13th Signal Regiment consists of about 250 experts who undertake the responsibility of formulating rules for, testing and deploying next-generation national defense network & information security operation center and provide secure network services for all military communications.

---

[9] https://xueqiu.com/6817277373/135373540.

[10] https://finance.sina.com.cn/roll/2019-08-10/doc-ihytcitm8141679.shtml.

[11] Data Source: British Department for Digit, Culture, Media and Sport (DCMS) released the report, January 30, 2020; Inspirations of British Cybersecurity Industrial Development to China, CCID, March 5, 2020.

### 1.3.4 *Singapore Strives to Build the World's Leading Digital Nation*

In Global Internet Development Index, Singapore ranks the 5th. In specific, Singapore ranks the 1st in information infrastructure index, the 13th in innovation capacity index, the 5th in industrial development index, the 26th in the Internet application index, the 7th in cybersecurity index and the 11th in cyberspace governance index. See Fig. 1.4 for Internet Development Index of Singapore.

Singapore keeps up with the pace of global IT development and constantly improves information infrastructure construction. Singapore achieves a high level of broadband coverage, especially the remarkable performance in the universalization rates of mobile broadband and smartphone. By June 2020, Singapore has issued two national 5G licenses. To make full use of 5G network, Singapore government has taken a series of measures, including launching the package storage cabinet network throughout the city. Singapore IMDA has announced that Singapore will launch two independent 5G networks from January 2021, build a 5G network that covers half of Singapore by the end of 2022, and build a 5G network that covers the entire Singapore by the end of 2025.

Singapore performs well in the environment of industrial development and the economic effect of digital industry and maintains a high level of ICT product export ratio and mobile application number. As the most attractive data center hub in the Asia–Pacific Region, Singapore can effectively meet the strong demand for high-end technologies such as cloud computing, big data and Internet of Things. Relying on the advantages in ICT industry, Singapore has vigorously implemented the blueprint plan for industrial transformation, promoted digital transformation, and actively accelerated the development of digital trade. In April 2020, Enterprise Singapore issued Small and Medium-sized Enterprises E-commerce Promotion Plan.

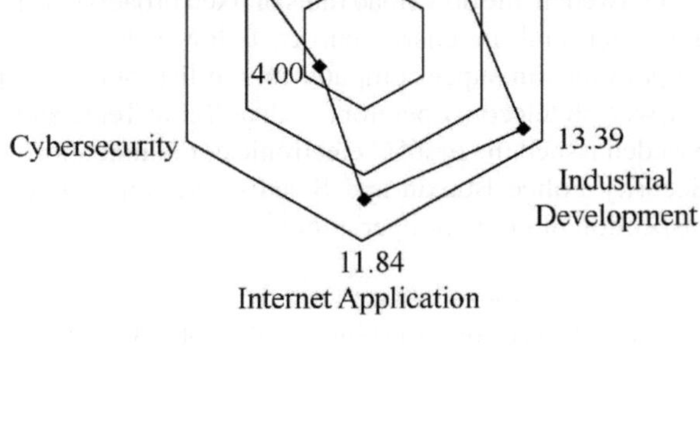


**Fig. 1.4** Internet development index of Singapore

In June 2020, Singapore signed digital economy agreements with Chile and New Zealand. In August 2020, Singapore signed the digital economy agreement with Australia, including avoiding unnecessary restrictions on data transmission and location, improving the protection of software source code and ensuring the compatibility of e-invoices and e-payment framework.

As Singapore maintains the first class in the quality of higher education in the world, it owns advantages in talents with digital skills and the market-oriented development of innovation outcomes. Yet, it needs to strengthen the number of ICT patent application and R&D investment.

Singapore performs well in business-level and government-level applications. Particularly, in recent years, Singapore government has constantly adjusted the e-government development plan and provided strong support for government-civilian-enterprise cooperation and innovation. As The United Nations E-Government Survey 2020 suggests, Singapore ranks the 11th in e-government development index in the world and lies in the "very high" level.

Singapore implements cyber content governance in multiple ways: establishing a "light touch" management framework, encouraging industrial self-discipline, and promoting public media quality and security awareness through public education. Meanwhile, Singapore continues to release new regulatory laws. In May 2019, Singapore Parliament passed the Protection from Online Falsehoods and Manipulation Act, which aimed to protect domestic networks from the harm cause by malicious actors who made lies and manipulated them. In 2020, Singapore plans to introduce data portability right and data innovation provision into the Personal Data Protection Act 2012 and strengthen personal data protection measures taken by public sector.

### *1.3.5 Sweden Displays Excellent Capacity in Internet Innovation*

In Global Internet Development Index, Sweden ranks the 6th. In specific, Sweden ranks the 10th in information infrastructure index, the 5th in innovation capacity index, the 6th in industrial development index, the 7th in the Internet application index, the 17th in cybersecurity index and the 12th in cyberspace governance index. See Fig. 1.5 for Internet Development Index of Sweden.

In Sweden, the download rates of fixed broadband and mobile broadband networks are faster, and the charge burden is lower. In Global 500, there are two Swedish corporations in super computer on the list. 5G develops rapidly in Sweden thanks to Swedish telecom operators such as Telia, Tele2 and Ericsson. In December 2019, Sweden issued the first 5G electronic communications law, which endowed Swedish Security Police Bureau and Swedish National Defense Force more power in the inspection of network operators.[12]

---

[12] Material Source: Svenska Dagbladet, December 30, 2019.

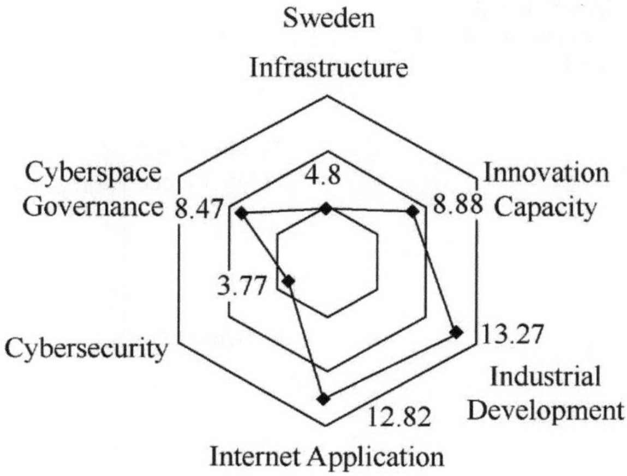


**Fig. 1.5** Internet development index of Sweden

Sweden enjoys the reputation of "Country of Innovation". It possesses advantages in talents with digital skills, R&D investment and marketization of innovation outcome. In Global Innovation Index 2019, Sweden ranks the 2nd in the world, only to Switzerland. Sweden continuously makes innovation in the Internet field. In February 2020, Sveriges Riksbank firstly tested e-krona, a digital currency that it released, to simulate daily banking activity.[13] In March 2020, Ericsson (Sweden) released native AI design in telecommunications, i.e., AI was put into the underlying architecture of telecommunication network to fully integrate AI into the set of network product and service.[14]

Sweden embodies high level of business-level and government-level applications in the Internet. In e-government, according to The United Nations E-Government Survey 2020, Sweden ranks the 7th in e-government development index in the world. Influenced by the COVID-19 Pandemic, e-commerce presents a rising trend. According to a survey, in the second quarter of 2020, 64% Swedish e-retailers believed that their sales increased.[15]

Sweden needs to vigorously promote the cybersecurity level, and no Swedish enterprise is listed in Global Hot Cybersecurity Enterprises top 150. In a survey on cybersecurity conducted by PricewaterhouseCoopers (Pwc) in 100 Swedish enterprises, 63% of respondents said that they suffered cyberattacks, up 13% year on year. Besides, 75% of respondents viewed that Sweden was not able to efficiently handle the growing cybersecurity threats, and as many as 96% of respondents hoped that courses on the cybersecurity education would be opened in primary and secondary schools.[16]

[13] Material Source: Reuters, February 21, 2020.

[14] Material Source: www.leiphone.com, March 25, 2020.

[15] https://finance.sina.com.cn/roll/2020-07-13/doc-iivhvpwx5177728.shtml.

[16] http://se.mofcom.gov.cn/article/jmxw/202006/20200602978199.shtml.

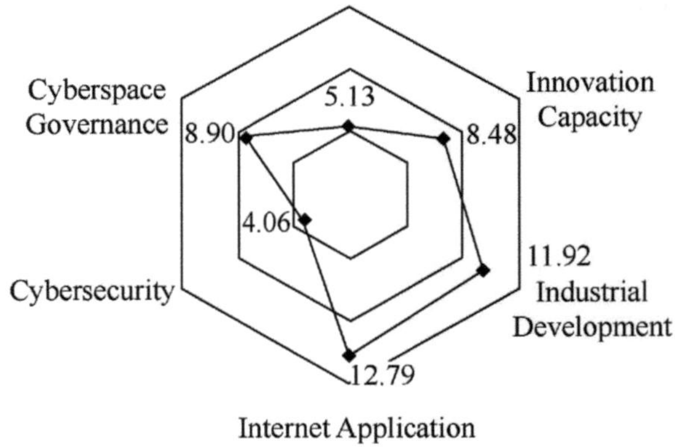


**Fig. 1.6** Internet development index of Canada

Sweden takes rigid rules on cyberspace governance. Swedish Data Protection Agency (DPA) strictly executes General Data Protection Regulation (GDPR) formulated by EU. In March 2020, Google was fined about 7 million euros as it violated the rule on the right to delete.[17]

### *1.3.6 Canada Possesses Prominent Advantages in Digital Governance and Cybersecurity*

In Global Internet Development Index, Canada ranks the 9th. In specific, Canada ranks the 6th in information infrastructure index, the 12th in innovation capacity index, the 16th in industrial development index, the 8th in the Internet application index, the 5th in cybersecurity index and the 7th together in cyberspace governance index. See Fig. 1.6 for Internet Development Index of Canada.

Canada has been committed to popularizing the Internet, with complete Internet infrastructure and nearly full coverage of broadband network across the country. In Canada, mobile Internet remains a more advanced level, and the download rate of mobile broadband network ranks the top in the world. According to the statistical data of Speedtest, by April 2020, the download rate of mobile broadband network reached 73.52 Mb/s in Canada, ranking the 6th among 139 countries and regions with statistics. In the assessment of 2019 Networked Readiness Index, Canada scored 74.72 in NRI and ranked the 14th in the world, the same as 2016.

Canada boasts distinct advantages in talents with digital skills, yet needs to strengthen R&D investment and the number of ICT patent application. Canada is the first country to launch AI strategy in the world. With the strong support of government, Canada has become one of global leaders in AI field. In particular, Montreal

---

[17] https://kuaibao.qq.com/s/20200322AZPNKU00?refer=spider.

has significant advantage in AI industrial clusters, especially in game and Virtual Reality (VR) technologies and industries. In Toronto-Waterloo region, AI industry comes to rise. Quantum technology is also a key field in Canada's investment. Canada owns two world-class quantum computing research centers: Perimeter Institute for Theoretical Physics and Institute of Quantum Computing (IQC) at University of Waterloo.

Canada takes the lead in the number of secure network servers per million people, and five cybersecurity corporations are listed in global top 150, second only to the United States, Israel and the United Kingdom. In Canada's National Cybersecurity Strategy (Revised Edition), it was clearly stated that "Strong cybersecurity played an essential role in Canada's innovation and prosperity". In 2019, Canadian Space Agency (CSA) granted Honeywell a $30 million contract, for the design and implementation of QEYSSat mission on quantum encryption and scientific satellite, to help improve Canada's cybersecurity level. Supported by National Cybersecurity Strategy (Revised Edition), Canada National Research Council (NRC) collaborated with University of New Brunswick (UNB) and established a new cybersecurity innovation center in Fredericton. The center will research cybersecurity for critical infrastructure and put emphasis on IoT security, AI, Human–Computer Interaction (HCI) and Natural Language Processing (NLP). In December 2019, Royal Canadian Mounted Police (RCMP) announced a five-year strategic plan to promote the establishment of a sound cybercrime law enforcement system in all regions. Canada has also launched a cybersecurity certification plan, requiring that small and medium-sized enterprises to abide by the baseline cybersecurity control measures established by Canada Cybersecurity Center.

In 2020, Canada Centre for International Governance Innovation and other institutions published several reports that explored the establishment and improvement of a sound digital trade regulatory system, proposed the establishment of a task group on data standards in charge of formulating global data cooperation standards, and advocated the establishment of a digital Bretton Woods System to coordinate global digital governance and reduce the negative impact of digital revolution.

### *1.3.7 Japan Accelerates the Development of Mobile Internet and Digital Economy*

In Global Internet Development Index, Japan ranks the 13th. In specific, Japan ranks the 15th in information infrastructure index, the 3rd in innovation capacity index, the 17th in industrial development index, the 13th in the Internet application index, the 17th in cybersecurity index and the 2nd together in cyberspace governance index. See Fig. 1.7 for Internet Development Index of Japan.

In infrastructure, Fugaku, a Japanese super computer, runs fastest in the world. Yet, there is a big gap in the number of Super Computer 500 between Japan and China/the United States. Besides, Japan needs to increase the download rates of fixed broadband

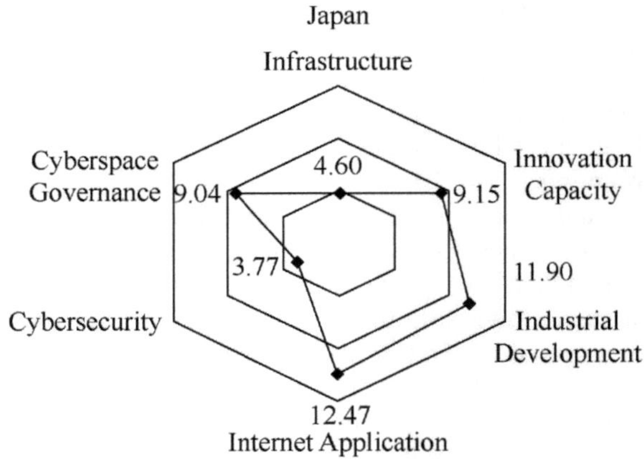


**Fig. 1.7** Internet development index of Japan

and mobile broadband networks. According to 2019 Networked Readiness Index, Japan's NRI dropped from the 10th (in 2016) to the 12th, which mainly resulted from that it ranked lower in the Internet access and content. To strengthen Internet access, Japanese operators accelerate the launch of commercial uses of 5G. In March 2020, Japanese corporation Softbank launched native commercial uses of 5G and became the first Japanese operator to provide services on commercial uses of 5G. It plans to deploy more than 10,000 5G network base stations by the end of 2023.

Owning to its prominent performances in R&D investment and the marketization of innovation outcomes, Japan has distinct advantages in innovation. According to 2019 EU Industrial R&D Scoreboard published by the European Commission, Japan was one of the countries/regions that boasted the largest number of corporations on the list. Specifically, Japanese corporations ranked the 3rd in R&D investment in the world (13.3% of the world), the 2nd in average R&D investment in the world (3.5% of the world), and the 3rd in average profitability in the world (7.8% of the world). In developing information process technology that supports 6G, Japan is building world standards. In April 2020, Japan's Ministry of Internal Affairs and Communications (MIC) issued the Outline for the Plan of 6G Integrated Strategy, stating that Japan would promote the research and development of 6G technology by means of financial support and preferential tax to achieve breakthroughs in key technologies in five years. In addition, Japanese Nippon Telegraph & Telephone (NTT) strives to establish the basis for information flow in the 2030s and proposes IOWN (Innovative Optical&Wireless Network) Vision. That is, basis for information flow such as semiconductor, personal computer, server and transmission system will be entirely borne by light.

Japan continues to quicken Internet applications, with high level of e-government development. According to The United Nations E-Government Survey 2020, Japan ranks the 14th in e-government development index in the world. Japanese prime minister Abe Shinzo expressed that in the next three years, Japan would centralize the investment and allocate budgets in key fields to promote the digital construction in

Japanese administrative field; in the next five years, Japan would establish a sharing service mechanism to facilitate and improve citizen's life.

In the industrial development of Internet, there is a large gap between Japan and the leading countries, so Japanese government keeps supporting industrial development of Internet. For instance, to drive Japanese corporations to develop secure 5G mobile network and unmanned aerial vehicles technologies, Japanese Cabinet passed a new act in February 2020. It allowed corporations that engaged in developing 5G and unmanned aerial vehicles (UAV) technologies to obtain low-interest loans from government-owned financial institutions on condition that their plans met the cybersecurity standards. According to the act, corporations that adopted 5G technology could be provided with preferential tax if they satisfied the standards Japanese government set up. In July 2020, Japan's Ministry of Economy, Trade and Industry released the White Paper on Trade 2020, emphasizing that Japan would enhance the investment into digitalization to adapt to the development in the post-COVID-19 Pandemic age. In 2020, Bank of Japan accelerated the research on digital currency. In January 2020, Japan jointly evaluated the feasibility of issuing Central Bank Digital Currency (CBDC) with central banks from the United Kingdom, euro zone, Sweden and Switzerland. In July 2020, Bank of Japan established the Digital Currency Division in Clearing Organization Bureau to concentrate on the research on Central Bank Digital Currency and explore the best clearing system in the construction of digital society. In Report on Technological Barriers in Central Bank Digital Currency published in August 2020, Bank of Japan disclosed that it would test the feasibility of digital currency technologically and continue to collaborate with other countries' central banks.

Japan has accumulated much experience in cyberspace governance. In recent years, Japan has continued to promote the free flow of data under safe conditions. In July 2018, Japan and EU signed Economic Partnership Agreement (EPA) and Strategic Partnership Agreement (SPA), hoping to jointly establish the "largest safe data flow area in the world". Afterwards, in January 2019, EU passed an "Adequacy Decision", which allowed the free flow of personal data between EU and Japan on condition that the data would be effectively protected. The decision complemented EPA and would come into force in February 2019. When EPA has been carried out for two years (since February 2021), Japan and EU will conduct the first joint review to assess the operation of the framework. Driven by Japan, G20 launched "Osaka Framework", which stressed that open cross-border data flow meant the lifeblood to all industries and that the strong protection of privacy and cybersecurity closely related to the transparency and non-discrimination of cross-border data flow. Accordingly, the idea of DFFT (Data Free Flow with Trust) was proposed. At the G20 Summit 2020, it is expected to further promote the free flow of data.

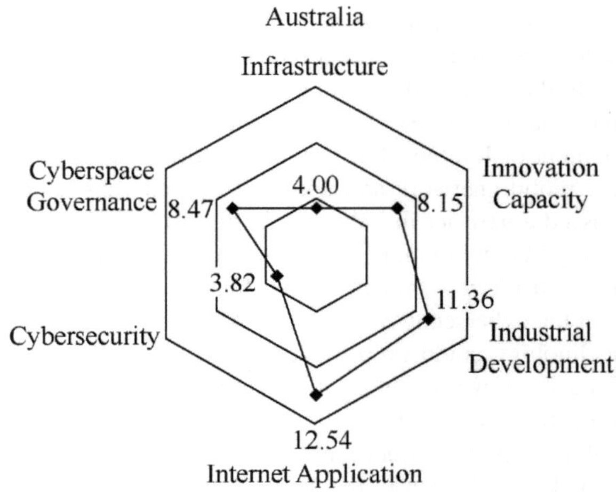


**Fig. 1.8** Internet development index of Australia

### *1.3.8 Australia Takes the Lead in the Internet Universalization Rate in the World*

In Global Internet Development Index, Australia ranks the 16th. In specific, Australia ranks the 22nd in information infrastructure index, the 15th in innovation capacity index, the 22nd in industrial development index, the 12th in the Internet application index, the 14th in cybersecurity index and the 12th together in cyberspace governance index. See Fig. 1.8 for Internet Development Index of Australia.

Australia performs well in IPV6 construction, the download rate of mobile broadband network and charge burden of mobile network. According to 2019 Networked Readiness Index, Australia's NRI scored 74.8 and ranked from the 18th (in 2016) to the 13th. This mainly resulted from the higher scores in the Internet access, personal skills, Internet security, Internet content and Internet governance.

Australia has deployed National Broadband Network (NBN) across the county and connected 3.91 million households and businesses places. To further quicken the network speed and improve the network service, all states in Australia start to build their own fiber optic networks. For example, New South Wales invests A$100 million (approximately $66 million) to implement Gig State Plan. Adelaide of South Australia continues to construct gigabit network in commercial areas.[18] In 2019, Australian e-commerce sales reached $32 billion, an increase of 17.5% over 2018. In the COVID-19 Pandemic, Australian e-commerce market developed rapidly. By the end of 2020, e-commerce market sales in Australia will account for 15% of the total domestic retail market.[19]

To promote the capacity of cybersecurity defense, Australian government announces that it will invest A$1.35 billion (approximately ¥6.6 billion) into the

---

[18] http://m.cableabc.com/world/20200525469520.html.

[19] https://www.egainnews.com/article/4792.

**Fig. 1.9** Internet development index of Italy

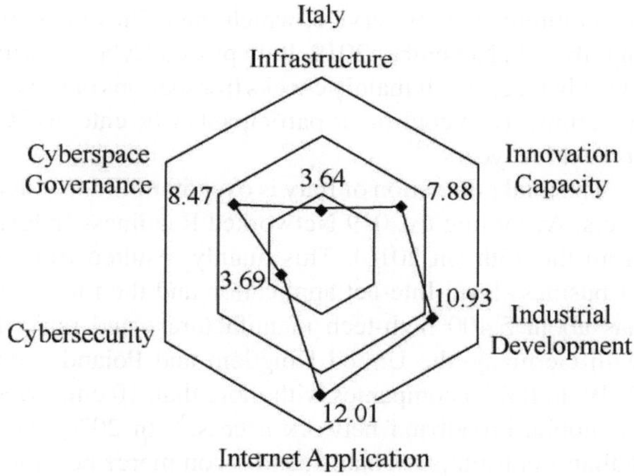


field of cybersecurity in next 10 years, of which A$470 million (approximately ¥2.3 billion) will be used to hire 500 cybersecurity experts.[20] To protect the security of personal data, in August 2019, Australia passed Consumer Data Right Act which stipulated the provisions on privacy and information security. This gave consumers independent rights to decide their data and choose between various products and services according to their conditions so as to better use their own data.[21]

Australia further strengthens cyber content management by legislative means. In April 2019, Australia's Criminal Law Amendment Act ruled that if Internet hosting service platforms failed to delete abhorrent violence content in a timely manner, the persons in charge of the hosting service would face a maximum of three years' imprisonment and a fine of 10% of their annual incomes.

### *1.3.9 Italy Maintains High Level in the Internet R&D Expenditure*

In Global Internet Development Index, Italy ranks the 23th. In specific, Italy ranks the 30th in information infrastructure index, the 17th in innovation capacity index, the 29th in industrial development index, the 22nd together in the Internet application index, the 23rd in cybersecurity index and the 12th together in cyberspace governance index. See Fig. 1.9 for Internet Development Index of Italy.

Italy is the 3rd country to achieve commercial uses of 5G, after Switzerland and the United Kingdom. In 2019, Vodafone opened 5G networks in five cities, Milan, Rome, Turin, Bologna and Naples. In 2020, it will cover 45–50 cities, and in 2021, it will cover 100 cities. In addition to Vodafone, Telecom Italia and Fastweb actively launch

[20] https://new.qq.com/omn/20200630/20200630A0KAL400.html.

[21] https://www.secrss.com/articles/13063.

5G communications service, which intensifies the competition on 5G construction in Italy.[22] In November 2019, Italy passed Cybersecurity Law to ensure the domestic 5G cybersecurity. It mainly checks transactions or uses the right to veto, so that Italian government can control the participants who enter its 5G telecommunications market in next few years.[23]

The total population of Italy is over 59 million, with more than 49 million Internet users. According to 2019 Networked Readiness Index, Italy's NRI ranked the 34th from the 45th (in 2016). This mainly resulted from the continuous improvement of business-level Internet application and the impact of Internet on economy. Italy has about 5,400 high-tech manufacturers and ranks top 4 in Europe, in parallel with Germany, the United Kingdom and Poland. According to data of ISTAT, in 2019, in Italian companies with more than 10 employees, 94.5% of them used fixed or mobile broadband network access.[24] In 2020, the COVID-19 Pandemic alters Italian's consumption habits, and e-commerce becomes the main engine to promote consumption. According to the survey of Alvarez & Marsal, Italy was once one of the countries with low universalization rate of online consumption. In 2019, the rate was 6.3%, and in 2020, it will reach 8.3%.[25]

Italy has invested enormously into scientific & technological research to promote the domestic Internet innovation capacity. According to Digital Economy and Society Index (DESI) 2020 published by European Commission, Italian enterprises expend an average of 2.4 billion euros in ICT R&D annually, which ranks the 4th in EU and follows after France (7.7 billion euros), Germany (6.5 billion euros) and the United Kingdom (3.7 billion euros). Italy stands above the European average level in the production and use of industrial robots, as well as the use of technologies such as cloud computing, Internet of Things and machine to machine communications. In particular, Italy ranks top 10 in robot density[26] in the world. According to the World Robotics Report 2019: Industrial Robot released by the International Federation of Robotics, Italy has 200 robots for every 10,000 manufacturing workers, far ahead of 168 in Spain and 154 in France.[27]

Italian data protection & regulation agencies are responsible for the domestic Internet regulation, including the supervision of data collection and processing. In July 2020, Wind Tre, an Italian telecom operator, was fined about 17 million euros as it failed to fully protect data and the rights of data subjects. Iliad, another telecom operator, was fined 0.8 million euros for traffic data collection.[28]

[22] https://tech.qq.com/a/20190607/000513.htm.

[23] http://it.mofcom.gov.cn/article/jmxw/201911/20191102914026.shtml.

[24] http://it.mofcom.gov.cn/article/jmxw/201912/20191202921059.shtml.

[25] http://it.mofcom.gov.cn/article/jmxw/202007/20200702980049.shtml.

[26] Robot density refers to the ratio of the number of industrial robots to the scale of labor.

[27] https://cloud.tencent.com/developer/news/447307.

[28] https://www.secrss.com/articles/24328.

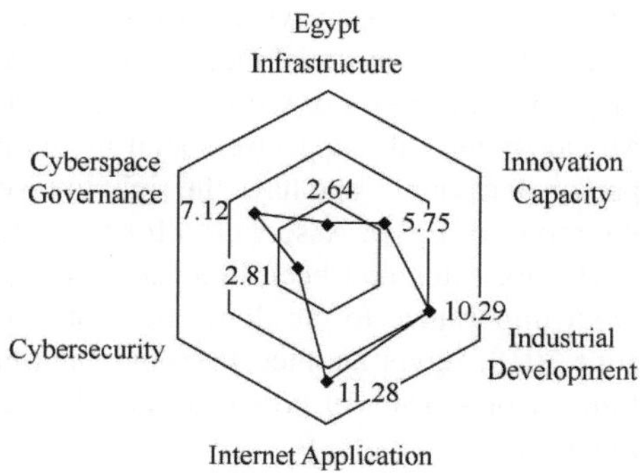


**Fig. 1.10** Internet development index of Egypt

#### 1.3.9.1 Egypt Displays a Rising Trend in the Internet Application

In Global Internet Development Index, Egypt ranks the 41st. In specific, Egypt ranks the 42nd in information infrastructure index, the 43rd in innovation capacity index, the 41st in industrial development index, the 40th in the Internet application index, the 47th together in cybersecurity index and the 42nd in cyberspace governance index. See Fig. 1.10 for Internet Development Index of Egypt.

Egypt has great potential in the development of network download rate, network charge, super computer construction and so on. According to 2019 Networked Readiness Index, Egypt's NRI scored 38.58 and ranked from the 96th (in 2016) to the 92nd. This mainly resulted from the higher scores in the Internet access, government-level Internet application, cybersecurity and the impact of Internet on economy.

Internet application in Egypt keeps rising. A survey by Egypt Institute of National Plan revealed that in early 2020, before the outbreak and spread of the COVID-19 Pandemic, only 8% of Internet users chose to shop online. During the COVID-19 Pandemic, online consumption data grew fast, with e-commerce sales expected to increase by more than 50%.[29] The usage of online education also rises rapidly. Egypt Ministry of Education has designated Edmodo, an online learning platform developed by China Online Education Company as the provider of distance education support for 22 million students and more than one million teachers in Egypt. In June 2020, Egypt Ministry of Communications and Information Technology and Huawei (China) jointly built AI classrooms at Aswan University, which provided technological training to Aswan and other provinces lack of educational resources.[30]

Egypt takes strict attitude in the Internet governance. It makes some achievements in the domestic cyberspace governance, yet needs to play greater role in international cooperation on cyberspace governance. Egypt attaches special importance to the administration of such illegal acts as rumormongering & slander and network infringement. In July 2018, Parliament of Egypt passed a bill on media regulation, empowering government to strengthen the management on social media users and to

[29] http://www.ce.cn/xwzx/gnsz/gdxw/202007/13/t20200713_35311516.shtml.

[30] http://www.ce.cn/xwzx/gnsz/gdxw/202007/13/t20200713_35311516.shtml.

crack down on false news. According to the bill, any accounts and blogs with more than 5,000 fans on social media such as Twitter and Facebook will be prosecuted if they post false news or stir up to illegal acts. Egypt Supreme Committee for Media Management will supervise social media platforms including websites, blogs and personal accounts, and have the right to take action against illegal acts. According to the report of Egypt Association for Freedom of Thought and Expression, by March 2019, 500 sites had been blocked in Egypt.[31] Egyptian government also attaches great importance to the domestic protection of personal data and information. In June 2019, Egypt adopted Personal Data Protection Law, which stipulated that the transfer or sharing of personal data abroad was prohibited and that violators were fined.

## 1.4 Trends of World Internet Development

Globally speaking, world Internet development enters a stag of new reform when we look into the future. Under the continuous influence of the COVID-19 Pandemic, Internet plays a more prominent role, and the number of global Internet users continues to rise. New-generation information technology represented by artificial intelligence, 5G and Internet of Things keeps innovating and developing, and further impacts and transforms social development and production & life models. Digital economy will accelerate its integration with real economy and provide new driving force for the development of world economy. Internet media content will continuously diversify and promote the interactive development and exchange of global cultures on Internet. Cybersecurity threats remain prominent, and all countries will continually improve their cybersecurity defense capacity. The changes of International landscape will further affect the Internet and intensify the uncertainty and vulnerability of international cyberspace governance.

In the future, all countries in the world should adhere to the connectivity of Internet, stand upon the base of a community with a shared future for mankind, uphold the concept of a community with a shared future in cyberspace, jointly promote the healthy development of Internet with cooperative and inclusive attitudes, strengthen the cooperation and innovation in the Internet technology, and jointly handle the risks and challenges of cyberspace so that the world can better use and enjoy the Internet.

---

[31] https://apnews.com/1540f1133267485db356db1e58db985b.

### 1.4.1 *Under the Circumstance of the COVID-19 Pandemic, the Degree of Dependence on Internet Increases, and All Countries Constantly Strengthen Information Infrastructure Construction*

When the COVID-19 Pandemic spread globally, telemedicine, online education, sharing platform, cooperative office, cross-border e-commerce and other services were widely used, which played an important role in ensuring social stability, promoting economic stability and enhancing international anti-epidemic cooperation. With the surge of demand for Internet from the whole society, the status and value of information infrastructure construction will prove more prominent and become a new development engine for all countries to vie for new technological dominant position, build new economic advantages and control the discursive power in global development. All countries in the world continuously increase their investment into the construction and R&D of new-type information infrastructure, take it as the key domestic strategic development direction at the current stage, and accelerate and empower the transformation and upgrading of economy, society and other fields or industries. In the future, in various countries, full coverage of 5G network will be achieved gradually, more devices will be connected to Internet of Things, more data centerss will be built, and Satellite Internet will be gradually put into use.

### 1.4.2 *The COVID-19 Pandemic Stimulates the Demand for Cloud-Terminal Coordination and Intelligentization, and Internet-Related Industries Face a New Development Opportunity*

The COVID-19 Pandemic activates the office & operation model that takes cloud-terminal coordination as the core and becomes an important complement to the mainstream office & operation model in various industries, which will produce the rise of new demand in the Internet-related industries like information technology. In various industries, cloud-terminal work & operation model that relies on advanced information technology such as cooperative office and remote control gives rise to explosive growth in demand. This also improves the information level of office & operation model in various industries in the future to some extent. In addition, it will stimulate the demand and development of sub-sectors of electronic information manufacturing industry such as integrated circuits, communication devices, computers, intelligent terminals and servers, and create more diverse use cases and demands for software & information service industry such as cloud computing, big data and artificial intelligence.

### 1.4.3 *Global Digital Economy Faces Great Uncertainty, and Global Internet Technological Ecology Displays a Diverse Trend*

Digital economy drives the real economy and connects the global market through the Internet. Besides, new business models and forms and industries that developed fast in the COVID-19 Pandemic shape new growth point of global economy. Yet, in terms of prospects, global economy faces great uncertainty, with both upside and downside risks. Given that the COVID-19 Pandemic will bounce back probably, restrictive measures on trade and investment increase, geopolitical uncertainty rises, and the future development of global digital economy displays an uncertain trend.

Meanwhile, international supply chain becomes more vulnerable under the influence of trade protectionism and unilateralism policies of individual countries. Some countries with high economic dependence on foreign countries start to reflect their industrial policies. In order to reduce excessive external dependence, some countries try to build their own technological ecology. This results in that the threats from the division in global Internet technological ecology are significantly intensified and that global cooperation on scientific & technological innovation is seriously hindered.

Global division and cooperation become more closely. Under such circumstance, to speed up the recovery of global economy, all countries should strengthen cooperation, explore local rules in line with trading reality and multi-interests, build an open, transparent, inclusive and non-discriminatory industrial development ecology and stable and secure global supply chain and industrial chain, and enlarge international cooperation in digital field in the form of bilateral and multilateral agreements.

### 1.4.4 *The Imbalanced Digital Development in Various Countries Probably Further Widens the World Digital Divide, and the Demand for Global Sharing of Digital Advantages Becomes More Urgent*

The popularization of Internet and the deepening of social digitization give development dividends to all countries in the world; however, there are huge gaps in the level and speed of digitization between various countries and within a country. For instance, in terms of digital readiness, developing countries lag far behind developed countries in many fields, such as ICT infrastructure, digital payment solution, skill and legal framework.[32] The imbalanced digital development in various countries probably further enlarges the world digital divide.

[32] Material Source: The COVID-19 Crisis: Accentuating the Need to Bridge Digital Divides, published at United Nations Conference on Trade and Development. See https://unctad.org/en/pages/PublicationWebflyer.aspx? publicationid = 2701, April 2020.

According to the survey of International Telecommunication Union (ITU), overcharging and lack of digital skills remain major constraints on the popularity and effective use of Internet, especially in least developed countries.[33] It constitutes an important direction in bridging the world digital divide on how to take measures to improve citizen's digital skills in underdeveloped countries with less developed Internet and to solve the problem of high prices of information devices and services as well. In the future, all countries face the great challenge to narrow the digital gap and quicken the sharing of digital dividends. This necessitates accelerating international cooperation in the digital field, jointly activating the innovation-driven cooperation momentum, and working together to create mutually beneficial and win–win cooperation among all countries.

### *1.4.5 Internet Omni-Media Displays a Distinct Trend, and Content Delivery Develops Toward a Refined State*

New-generation information technologies represented by 5G, artificial intelligence and Internet of Things expands the use cases of Internet and seeks to establish business model, which further broadens Internet media market. Traditional media speeds up the transformation into digital media, video streaming media develops rapidly, and Internet media become platforms with a more evident trend. With the continuous development of media integration and the application of new technology, omni-media platform with various functions such as payment, interaction, entertainment and consultation will become one of the trends in the Internet media development in the future.

Meanwhile, as the technological application of artificial intelligence and big data enlarges, content production and delivery models on information will be more refined, and information interaction for different users and use cases will be more common and intelligent. However, as more cyber information is released, false and malicious information also increases, which seriously interferes with the order of Internet communication and calls on governments, enterprises and Internet users to participate in governance.

---

[33] Material Source: "Globally there are 52% of women inaccessible to Internet with enlarged gender digital divide" published by ITU.
See https://news.un.org/zh/story/2019/11/1044991.

### 1.4.6 The Militarization and Competition of Cyberspace Intensify, and the World's Cybersecurity Situation Becomes More Serious

Cyberspace plays an increasing important role. As various countries enhance their cybersecurity defense capacities, the militarization trend of cyberspace becomes more evident and poses great challenge to the security and stability of cyberspace. In addition, all countries in the world face the issues on how to ensure the securities in cyberspace and communication and to prevent the network monitoring. As The Washington Post (the United States) reports, the Unite States Central Intelligence Agency (CIA) has been monitoring confidential communications in hundreds of countries in the world for decades of years by inserting encryption holes in products sold by Crypto AG, a Swiss encrypted communication device corporation.

The situation of global cybersecurity remains severe. Problems on cybersecurity such as cyberattacks, data leakage and network virus occur frequently at different places. In particular, in the COVID-19 Pandemic, the wide use of online industry and telemedicine will exacerbate global cybersecurity risks, and all countries still concentrate on network data and privacy protection.

### 1.4.7 Global Cyberspace Landscape Accelerates Its Change, and International Internet Governance Model Faces Adjustment

The deep adjustment of international landscape and the world order poses new challenges and opportunities to cyberspace governance. In international landscape, the trend of multi-polarization becomes prominent and the collaboration on promoting cyberspace governance becomes more important among various countries. All countries strengthen support for cyberspace development and enhance the governance of cybersecurity, data flow and other fields, which establishes a foundation for their jointly improving the global cyberspace order. The COVID-19 Pandemic will profoundly alter people's work and life styles and give birth to a series of new cyberspace forms and governance means. Under the background of the COVID-19 Pandemic, technological innovation will arouse a new round of scientific & technological revolution and give new technological support to cyberspace governance. Global cyberspace landscape will accelerate its change.

In the future, the deep adjustment of international landscape will be superimposed on the impact of the COVID-19 Pandemic, and cyberspace governance model will be probably changed. State actors will play a more remarkable role, and Internet communication will become normal among heads of states. Non-state actors represented by large Internet corporations will exert enlarging influence on cyberspace, and traditional international organizations need to strengthen their influence on cyberspace. Non-technological factors and various "Black Swan" incidents will deepen the reform on international cyberspace governance and the multi-stakeholder governance model needs to be explored and improved.

## 1.5 Joining Hands to Build a Community with a Shared Future in Cyberspace

Internet boasts one of the greatest inventions of human in the twentieth century. 51 years ago, human invented the Internet, which enormously expanded human's space for production and life, shortened the global distance, connected all countries in the world, and transformed international community into a community with a shared future where human were interdependent. Meanwhile, a new round of scientific & technological revolution and industrial transformation incubated and emerged, which caused the strong rise of digital technology and created the vigorous development of digital economy. Today, 51 years later, the world Internet enters a stage of fundamental change. The COVID-19 Pandemic prevails globally, economic globalization encounters adverse condition, and trade protectionism and unilateralism emerge. As a consequence, the world economy slows down, international trade and investment shrink sharply, global scientific & technological innovation and cooperation suffer disturbance, and non-economic factors destroy the global industrial chain and supply chain.

Various countries and international organizations in the world actively explore the new Internet governance model with multiple-party participants. For example, Secretary-General of the United Nations has established High-level Panel on Digital Cooperation in order to promote relevant recommendations and strengthen cooperation in digital space among governments, private sectors, civil institutions, international organizations, technological & academic communities and other stakeholders. Since 2014, China has held annually World Internet Conference, which creates an important platform for the connectivity across all countries in the world and international Internet sharing and co-governance, makes outstanding contribution to building a community with a shared future in cyberspace, and continuously enlarges consensus among all parties.

In spite that the deep adjustment of international landscape and the world order as well as the global spread of the COVID-19 Pandemic have led to uncertainty in the development of cyberspace, peace and development remain the theme of the times and people of all countries bear stronger expectations of peaceful development and win–win cooperation. International community needs to strengthen communication, build consensus, deepen cooperation, and jointly construct a community with a shared future in cyberspace in order to actively deal with the challenges and risks of cyberspace and explore the new world in the Internet.

To build a community with a shared future in cyberspace means to embody and practice the concept of a community with a shared future for mankind in a specific and important way, to construct cyberspace into a community of shared development, security, responsibility and interests that benefits all the mankind, and to make the inevitable choice that relates to human's future and fate and conforms to the development trend in the information age. On the basis of respecting each other's core interests, all countries should seek common well-being and handle common challenges so that Internet will better benefit people of all countries in the world!

# Chapter 2
# The Construction of World Information Infrastructure

## 2.1 Outline

Presently, human society is undergoing a new-round scientific & technological revolution and industrial transformation. New-type information infrastructure represented by 5G, AI, IoT and industrial Internet gradually become the new kinetic energy for global economic growth. All countries attach great importance to the development of new-generation technology, and accelerate and empower the industrial transformation and upgrading of various sectors in economy and society, in order to stay ahead in technological progress, build new economic strength and maintain the dominant role in global development.

1. The Evolution and Upgrading of Basic Network Accelerate

5G comes to a new stage of full-scale commercial use, major global telecom operators quicken the deployment of 5G network, and the scale of users grows rapidly. The upgrading and iteration of fix broadband network speed up. The number of global fixed broadband users grows slowly, yet that of optical fiber users maintains a trend of high growth. All countries continue to deepen universal service in telecommunications, hasten the construction of broadband networks in rural and remote areas, and further narrow the digital divide. The competition in the construction of spatial information infrastructure turns white-hot, the construction of satellite Internet and satellite IoT accelerates, and the development of global satellite navigation system presents a new landscape.

2. Application Infrastructure Develops Stably

The construction of intelligent infrastructure represented by data centers, AI and blockchain accelerates, global cloud computing market develops fast, AI HashRate and algorithm continuously improve, public infrastructure of blockchain faces a period of rapid development, and global IPv6 allocation continues to increase.



Chinese Academy of Cyberspace Studies, *World Internet Development Report 2020*,
https://doi.org/10.1007/978-981-16-9388-5_2

3. The Global Layout of New-Type Facilities Quickens

Major telecom operators represented by Vodafone, AT & T and China Telecom speed up the layout of Cellular IoT, and the number of IoT sensing terminal connection shows a trend of high growth. The core of the platform gradually changes from original Connectivity Management Platform (CMP) to Device Management Platform (DMP) and Application Enablement Platform (AEP). The platform is also transformed. The construction of global industrial IoT achieves positive progress, and the construction of core infrastructure (like the analysis system of network facility and logo) speeds up. The market scale of industrial Internet continually expands, whose driving capability for the digital transformation of manufacturing industry gradually emerges.

## 2.2 The Evolution and Upgrading of Basic Network Speed Up

The construction of global basic network facilities accelerates, the deployment of 5G network makes remarkable progress, and the scale of users keeps growing. The construction of high-speed broadband network quickens, and global universal service in telecommunications advances in an orderly way. In the COVID-19 Pandemic, broadband network played a prominent crucial supporting role. The construction of international submarine cable reaches its peak, focusing on the cross-Pacific submarine cable system. Globally, the construction of satellite Internet (comprised of high-orbit and low-orbit constellations) and low-orbit satellite IoT constantly advances, in which global satellite navigation systems integrate with each other in their competitive upgrading and iteration.

### *2.2.1 Global 5G Network Develops Rapidly*

1. The Deployment of 5G Network Fully Starts

As the data of Global Mobile Suppliers Association (GSA) suggests, by May 2020, global 80 telecom operators in 42 countries and regions had launched 5G commercial services that met 3GPP standards. Meanwhile, 384 telecom operators in the world had invested into 5G network, which are tested, examined, piloted, planned or deployed now. 95 telecom operators had announced the deployment of 5G technologies in their networks, which satisfied 3GPP standards. In terms of 5G business service, telecom operators can provide differentiated services according to such information as geographical location, terminal model, terminal type and user's grade.

2. The Scale of 5G Users Enlarges Rapidly

Mobile users are transferred to 5G network. According to *Ericsson Mobility Report*, by the end of 2020, the number of global 5G users would reach nearly 200 million and mainly increase in South Korea, United States and China. In April 2019, South Korea firstly started 5G commercial services in the world. Stimulated by terminal subsidy and charge subsidy, the number of 5G mobile users in South Korea grew by leaps and bounds. According to the data South Korean government published, by April 2020, the number of 5G users had reached 6.34 million, accounting for *c.* 10% of the total number of mobile users. In United States, the number of 5G users also keeps growing. According to the reporting data of M-Science, by July 2020, it had reached 4.082 million.

3. 5G Terminal Becomes More Diverse

5G terminal thrives, and the price of 5G mobile phone continues to fall. On the basis of *Global 5G Terminal Ecosystem Report* issued by the Global Mobile Suppliers Association (GSA), by May 2020, global 84 suppliers had released 296 5G devices. Among them, 5G mobile device and 5G Customer Premise Equipment (CPE) ranked the first and second, with 119 types and 84 types. In addition, the devices included 5G modules, laptops, drones, tablets and others. Globally, South Korea and China take the lead in the penetration rate of 5G mobile phone, and United States, Russia and European Union also increased in this regard. According to the reporting data of M-Science, by July 2020, a total of 4.1 million 5G mobile phones were sold in United States.

4. The Allocation of 5G Spectrum Was Actualized and Optimized

5G requires wider frequency band to support the higher speed and larger flow. The plan for allocating 5G spectrum for major countries in the world is gradually actualized. Under the spectrum planning framework of International Telecommunication Union (ITU), countries draw up their strategies for 5G spectrum based on their division and use of frequency. As the report of Global Mobile Suppliers Association (GSA) shows, the main 5G frequency bands are 700 MHz, 3400–3800 MHz and 24–29.5GGHz, and most global 5G networks are deployed in 3.5G MF frequency band. Federal Communications Commission (FCC) provides HF millimeter-wave frequency band that is used in the deployment of 5G commercial network. European Union uses 700 MHz, 3.4–3.8 GHz, 24.25–27.5 GHz, 31.8–33.4 GHz and 40.5–43.5 as 5G frequency bands. In the Asia–Pacific Region, three major South Korean operators use 3.5 GHz frequency band. And China has clearly divided 5G frequency bands, 3.4–3.5 GHz for China Telecom, 3.5–3.6 GHz for China Unicom, and 2515–2675 MHz, 4.8–4.9 GHz for China Mobile respectively.

### 2.2.2 4G Network Achieves Deep Coverage

1. The Construction of 4G Network Continuously Deepens

Globally, the number of operators with 4G network service continues to rise. According to the statistics of Global Mobile Suppliers Association (GSA), by the end of March 2020, 797 telecom operators in the world had launched Long Term Evolution (LTE) commercial network (including mobile + FWA). Among them, 401 had launched LTE fixed wireless access service, 729 had offered comprehensive mobile LTE service, 234 had held TD-LTE service spectrum licenses. In terms of technological deployment, 325 telecom operators in the world had deployed LTE-Advanced or LTE-Advanced Pro technologies in commercial network. In addition, 270 telecom operators are investing into VoLTE technology, among which 210 telecom operators can provide services.

2. The Number of 4G Users Grows Slowly

As 5G is extensively used in commerce, the scale of global mobile phone users grows steadily, the number of 4G users increased slowly, and 4G users flow quickly to 5G network in some countries. According to the data Global Mobile Suppliers Association (GSA) released, by March 2020, the number of global 4G users reached 4.199 billion, which accounted for 67.4% of the total users and rose by 0.3% over the end of 2019. China ranks the first in the world in the number of 4G users (1.29 billion). In Japan and France, the numbers of 4G users grow rapidly, by more than 10% over the corresponding period. In South Korea, negative growth occurred in the number of 4G users in 2019. In March 2020, the scale of 4G users decreased by 983, 000 compared with the end of 2018. See Fig. 2.1 for "Proportion of 4G Users in Major Countries in the World in 2013–2020 (March)".

### 2.2.3 The Construction of High-Speed Broadband Network Accelerates

1. The Number of Fixed Broadband Users Grows Slowly

The number of fixed broadband users grows slowly, or declines in some countries and regions. As the data of Point Topic suggests, by March 2020, the number of global fixed broadband users reached 1.13 billion, up 5.8% year on year, yet slightly down compared with the previous quarter (6.8%). The number of global FTTH users continued to grow, reaching 503 million households. In specific, China and Brazil ranked the first and the second in quarterly net increase (6.89 million and 1.238 million households respectively); Belgium and Thailand ranked the first and the second in quarterly growth rate (26.3% and 14.2% respectively). See Fig. 2.2 for "Growth of Global Fixed Broadband Users in 2013–2020 (March)".

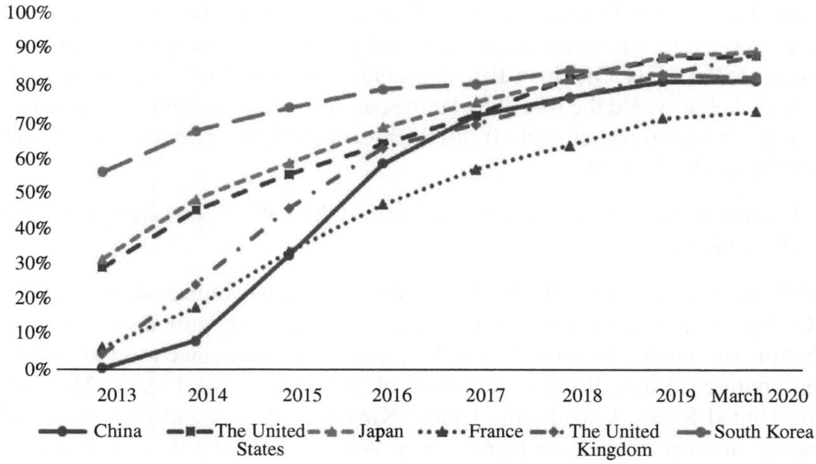


**Fig. 2.1** Proportion of 4G users in major countries in the world in 2013–2020 (March). (*Data Source* GSMA)

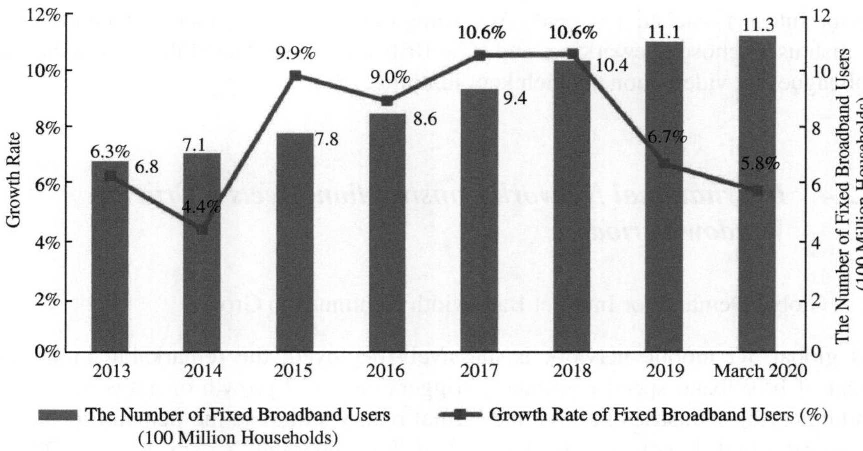


**Fig. 2.2** Growth of global fixed broadband users in 2013–2020 (March). (*Data Source* Point Topic)

2. Universal Service in Global Telecommunications Continually Advances

To further narrow the digital divide, all countries continuously deepen the universal service in telecommunications and issue related policies and acts to promote the construction of broadband network in rural areas. In March 2020, Danish Energy Agency drafted *2020 Broadband Subsidy Plan*, in which Danish government allocated DKK 100 million (*c*. $15 million) for the construction of broadband network in rural areas. In May 2020, United States issued *Universal Broadband Act*, which extended the scope of Universal Service Fund (USF) support to broadband service. It requested U.S. Federal Communications Commission to fix charge rate if necessary, provide service to all American citizens and prioritize the support to non-service areas and tribal areas. Ofcom, a British regulator, proposed to compensate telecom operators that provided the broadband universal service, and ordered BT and Kcom, two British telecom operators, to fulfill the broadband Universal Service Obligation (USO) from March 2020.

3. Broadband Network Powerfully Supported Office Work and Life in the COVID-19 Pandemic

In 2020, the spread of the COVID-19 Pandemic enormously impacted on the world. In the COVID-19 Pandemic, broadband network effectively supported socio-economic operation and citizen's production & life, and firmly guaranteed the teleworking, video conference and distance education of global broadband users. Meanwhile, in the United States, Canada, the United Kingdom, Russia and European Union, domestic Internet traffic rose significantly. According to the data of Comcast, in March 2020, the total Internet AV call traffic grew by 212%, up by 32% over the recorded peak traffic. As the data of Canadian government indicated, from mid-March to late April 2020, downloads and uploads from/onto the cable Internet increased by 48.7% and 69.2% respectively. In the United Kingdom, the times adults spent on the Internet reached new peak. According to the research data of Opinium, 69% British users chose teleworking, and 62% British users contacted their families and colleagues by video phone, which kept jumping.

### *2.2.4 International Network Construction Meets a Crucial Window Period*

1. Global Demand for Internet Bandwidth Continues to Grow

As global 5G mobile network is massively deployed, the remarkable improvement of broadband speed continually triggers the rapid growth of network traffic, and emerging industries like AI and virtual reality stimulate the demand for next-generation high bandwidth. In the light of TeleGeography's data, by June 2020, global Internet bandwidth reached 618 Tb/s, nearly three times of the similar data in the end of 2016. In Africa and Asia, the demand for intra-Internetwork bandwidth grows quickly. In 2016–2020 (June), Compound Annual Growth Rate (CAGR) of Internet bandwidth realized 46% and 40% respectively. As ever, global Internet traffic mainly flows to the United States, and the regional routing with the largest bandwidth capacity circulates around Latin America-United States. In 2020, the bandwidth of the regional routing increased by 30% and reached 56 Tb/s. See Fig. 2.3 for "Regional Compound Annual Growth Rate (CAGR) of Internet Bandwidth in 2016–2020 (June)".

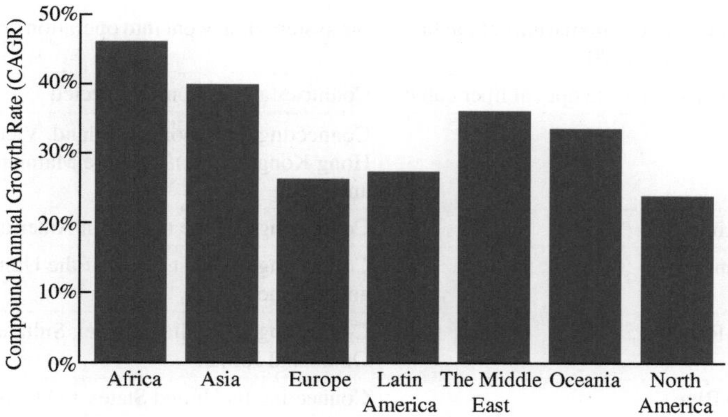


**Fig. 2.3** Regional compound annual growth rate (CAGR) of Internet bandwidth in 2016–2020 (June). (*Data Source* TeleGeography)

2. The Construction of International Submarine Optical Fiber Cable Reaches Its Peak

Globally speaking, the demand for international Internet bandwidth and the replacement cycle of submarine optical fiber cable co-contribute to the arrival of its new constructive peak. According to the data of TeleGeography, by January 2020, there were around 406 submarine optical fiber cables in use in the world, with a total length of over 1.2 million kilometers. In 2016–2020, 107 newly-constructed submarine optical fiber cables went into operation in succession, with a total length of more than 400,000 km. East Asia and North America boast two centers of the world Internet, and present an irresistible trend to connect them on equal terms. Simultaneously, owing to the open markets and well-equipped data centers clusters in China's Hong Kong, Los Angeles and San Francisco, submarine optical fiber cables that take these international free ports as landing point become new favorable choices for industrial investment accordingly. In terms of submarine optical fiber cable systems that went into operation or were under construction in 2019–2020 (see Table 2.1), the systems on two banks of Trans-Pacific remain the leading position. Besides, international submarine optical fiber cable continually extends to more countries. In recent years, several (island) countries, including Solomon Islands and the Cook Islands, got access to global Internet via submarine optical fiber cable.

**Table 2.1** Statistics of Submarine optical fiber cable systems that went into operation or were under construction in 2019–2020

| No | Name of submarine optical fiber cable | Countries and regions connected |
|---|---|---|
| 1 | SJC2 | Connecting Singapore, Thailand, Vietnam, Hong Kong (China), Chinese Mainland, Korea and Japan |
| 2 | EllaLink | Connecting Europe and Latin America |
| 3 | Dunant | Connecting the East Coast of the United States and Europe |
| 4 | Blue-Raman | Connecting India, Italy, Israel, Sultanate of Oman and Jordan |
| 5 | Deep Blue | Connecting theUnited States and Cuba |
| 6 | Curie | Connecting the United States and Chile |
| 7 | JGA-S | Connecting Chiba, Guam and Northern Australia |
| 8 | PEACE | Connecting Asia, Africa and Europe |
| 9 | Bay to Bay Express | Connecting Singapore, Hong Kong (China) and the United States |
| 10 | 2Africa | Connecting Africa, Europe and the Middle East |

### *2.2.5 Competition on the Construction of Spatial Information Infrastructure Turns White-Hot*

1. The Construction Satellite Internet Network Accelerates

Currently, major global satellite manufacturers and operators actively develop satellite Internet that consists of high-orbit satellites and low & medium-orbit constellations. High-orbit satellite develops stably, and traditional corporations play a leading role in manufacturing and launching satellites. Predictably, after 2020, the order for high-orbit satellites (GEO) covers about 10–15 every year. Presently, representative on-orbit GEO HTS are Viasat-2 from Viasat and Jupiter-2 from Hughes (the United States), whose capacities reach 300 Gb/s and 220 Gb/s respectively. Viasat-3 and Jupiter-3 are under construction. With the capacity of 1 Tb/s and 500 Gb/s respectively, they will be launched in 2021 and 2022. In medium-orbit constellation, O3b boasts the only medium-orbit broadband constellation system for commercial use in the world. 20 satellites of its Phase-I are all on orbit, and mPower, a 2G constellation, is under construction. Its capacity will tremendously improve after completion. In low-orbit constellation, the competition for satellite frequency and orbit becomes increasingly white-hot. SpaceX takes the lead. By June 2020, it had launched a total of 480 small satellites (eight batches). On condition that the scale of satellite reaches 800, Starlink constellation can basically provide services to the world.

2. The Development of Global Satellite IoT Speeds up

Low-Orbit Satellite IoT (LEO IoT) functions as an effective way to solve the insufficient coverage of the ground IoT in oceans, mountains, deserts and other areas. It uses low-orbit satellite constellation to realize IoT and owns such advantages as low delay, low transmission loss and global seamless coverage (including two poles). In particular, it can solve the problem of unsatisfactory communication in certain terrains. As low-orbit small satellite constellation grows rapidly, the development of global satellite IoT accelerates. In addition, many countries carry out Satellite IoT Program now. HeliosWire, a Canadian corporation, plans to launch 30 satellites to build spatial IoT, which uses 30 MHz bandwidth (S wave band) and supports the data collection of 5 billion sensors. Kepler Communications (the United States) plans to achieve the 3-Phase deployment of spatial network in 2023. It consists of 140 NanoSats that operate on 575 km-high polar orbits. Hongyan Communications Satellite Constellation System, constructed and operated by China Aerospace Science and Technology Corporation (CASC), serves as a global low-orbit satellite mobile communications & spatial Internet system. Initially, it consisted of more than 50 satellites. With over 300 small low-orbit satellites after completion, it can globally provide six types of applications and services: mobile communication, broadband Internet access, IoT, navigation enhancement, aviation data business and navigation data business.

3. The Development of Global Satellite Navigation System Presents New Pattern

The co-existence pattern takes shape in global satellite navigation multi-system. Compatibility and interoperability are unanimously accepted as a developmental trend. The new-round competition centers on providing diverse services and upgrading original systems. U.S. Satellite Navigation System (GPS) successfully launched the first GPS IIIA satellite in 2018. It plans to complete the deployment of 10 GPS IIIA satellites in 2023 and of 22 GPS IIIF satellites in 2034. Russian GLONASS plans to build a GLONASS-KM-centered satellite constellation by 2030 and launch 6 high-orbit GLONASS-B satellites by 2025, which will quicken network performance in Eastern Hemisphere by 25%. In June 2020, China successfully launched the 55th navigation satellite of BDS, which signaled the deployment of China's BD3 Global Satellite Navigation System Constellation was fully completed. In 2035, national comprehensive positioning navigation time service systems will be established, which take BeiDou as the core and use various means to realize the supplement, copy and enhance each other. European GALILEO plans to launch the third batch (four) FOC (Full Operational Capability) satellites in 2020–2024, launch 2G satellite in 2025, and form FOC for 2G system in around 2035.

## 2.3 Application-Oriented Infrastructure Develops Steadily

Global application-oriented infrastructure enters a crucial developmental stage. The scale of data centers grows stably, and international leading corporations on data centers intensify their global business expansion. Infrastructure on cloud computing and edge computing possess great developmental potential and vast market space. Various countries gradually lay strategic stress on new-type infrastructure & technology represented by AI and blockchain. The construction of global domain name market and facility advances steadily, and IPv6 commercial deployment fully develops, with stable growth in penetration rate and traffic.

### *2.3.1 The Number of Global Data Centers Grows Stably*

1. Global Data Centers Assemble in the Asia–Pacific Region

In 2020, the number of data centers increased the most in the Asia–Pacific Region, Europe, the Middle East and Africa. In Europe and the United States, as the market for traditional data-center business was basically saturated, large transnational corporations continuously constructed data centers in major regions in the world to enhance their capabilities for global service. According to the data of Synergy Research, the number of data centers in the Asia–Pacific Region accounted for more than 60% of the total number in the world, with the United States (40%) and China ranking the first and the second respectively. Japan also expanded quickly, which contributed to the fastest growth in data-center market in the Asia–Pacific Region. As newly emerging market economies, Singapore, Malaysia and India developed rapidly. They attracted the arrival of many transnational corporations with the excellent environment for IT industrial development, abundant supply of energy, preferential policies, complete laws and regulations and abundant resources of international submarine optical cable. This forged industrial agglomeration effect and greatly promoted the leapfrog development of national economy.

2. International Leading Corporations on data centers Boost Global Business Expansion

International leading corporations on data centers boost global business expansion. They establish data centers in the world and provide global services by investment and M&A (merger & acquisition). As the research report of Synergy Research suggests, in 2019, the number of M&A transactions on data centers firstly exceeded the 100 (cases), up by 6% over 2018 and over two times of that in 2016. In 2015–2019, the transactions of Digital Realty and Equinix, two leading hosting service providers in the world, took up 31% of the total. Digital Realty operated data centers in more than 30 markets in North America, Europe, South-East Asia and Australia. In 2019, it purchased Interxion, a European corporation, at the price of $8.4 billion, which was called the biggest data-center acquisition transaction in history. Equinix owns more than 200 data centers in five continents and covers 52 major markets. In 2019, it purchased AMS1 data centers of Switch (at Amsterdam, Holland) and three data centers in Mexico.

3. Cross-Border Data Policy Accelerates the Construction of Local Data Centers

In recent years, several countries carried out restrictive policies on the cross-border data flow, which resulted in the increase of local data centers. South Korea, Russia, Malaysia, Venezuela and Nigeria enacted related laws and regulations, stipulating that some IT infrastructure must be kept domestically, which aroused more demands for constructing national data centers. At present, the restrictions on the cross-border data flow are mainly executed in two ways. As required, firstly, data centers must be established domestically, and secondly, data must be stored domestically. In some countries, corporations are allowed to run business on condition that they establish data centers there. Vietnam demands that all Internet providers, including Google and Facebook, establish at least one data center internally. Indian government requires that corporations must establish some IT infrastructure in India so that inspection departments can get access to encrypted information.

### *2.3.2 The Infrastructure of Cloud Computing and Edge Computing Displays Enormous Potential in Development*

1. Global Competition on Cloud Computing Intensifies

Globally, competitive pattern continues in the market of cloud computing. AWS still dominates the global cloud infrastructure service market, yet its market share starts to decline. According to the latest research report of Canalys, in 2019, AWS ranked the first, with the market share of 32.3%; Azure (16.9%), Google Cloud (5.8%) and Ali Cloud (4.9%) followed after it. In China, Ali Cloud plays a decisive role in cloud computing and maintains fast growth (with a growth rate of 63.8%). It quickly narrows the gap with Google Cloud in market share. Presently, in cloud computing industry, three major competitive groups has initially formed, i.e. Internet, IT and telecom operators. In all three camps, in specific, Internet Group mainly concentrates on public cloud market and provide public-cloud service for small & medium-sized enterprises and independent developers. IT mainly focuses on key customers and provides private cloud products and solutions. Telecom operators engage in both public and private cloud markets, and provide IaaS public cloud service, governmental & industrial cloud hosting and customized services.

2. Global Edge Computing Market Owns Great Development Space

Edge computing has become a hot research topic in the fields of IoT, 5G mobile communications and cloud computing. Various industrial alliances, standardization institutions, open-source organizations and commercial enterprises actively promote the research, standardization and industrialization of edge computing. European and American enterprises dominate the global edge computing market, and major enterprises on cloud computing, chip and system integration successively launch their solutions on edge computing. In the Asia–Pacific Region, Japanese, South Korean and Chinese enterprises also actively foster the application of edge computing into communications, industrial development, health care and smart city, playing a positive role in R&D and industrialization of global edge computing. According to the report of Markets and Markets, the scale of global edge computing market reached $2.8 billion in 2019 and will grow to $9 billion in 2024 as predicted, with Compound Annual Growth Rate (CAGR) of 26.5%.

### *2.3.3 AI Platform Faces a Period of Rapid Development*

1. Major Countries Release Their AI Strategies

Globally, AI faces a new wave of development. Major countries extensively release AI strategy, strengthen the layout of AI infrastructure, and stress the investment into data and algorithm. In January 2019, European Union launched "AI FOR EU" Project, which aimed to establish an open and collaborative AI platform, integrate AI resources of European enterprises like data, computing, algorithm and tool, and provide uniform and open services. In June 2019 and February 2020, United States issued *National Artificial Intelligence Research and Development Strategic Plan* and *Federal Data Strategy and 2020 Action Plan* successively, proposing to develop public data set and environment for AI training and testing, and establish federal data resource library and standard database so as to assess the maturity of data infrastructure construction.

2. AI Platform Construction Achieves Remarkable Results

AI platform infrastructure is an open and innovative carrier for open source, which concentrates on key subdivided fields of AI, effectively integrates resources of technology, industrial chain and finance, and continuously generates core R&D and service capabilities of AI. Various countries speed up the layout of AI platform construction. The United States is market-oriented. Technological giants like Google, Amazon and Microsoft spontaneously invest into the establishment of AI open platform, and improve their industrial ecology. Meanwhile, U.S. government actively builds AI open platform for specific application fields. European Union strives to provide the world-class AI tool, component, module, knowledge, algorithm and case by constructing AI open platform that can be used universally, via the government-led model.

### 2.3.4 Blockchain Encounters a Period of Outbreak

1. Various Countries Return from Fictitious Economy to Real Economy in Their Blockchain Strategies

Countries in the world attach great attention to the application of blockchain into the field of real economy, and some clarify supervision policies on cryptocurrency. It is a general trend to return from fictitious economy to real economy in blockchain strategy. China, Holland, South Korea, Germany, the United States, the United Kingdom and Australia actively explore the R&D and application of blockchain technology. In September 2019, Germany released *Blockchain-Strategie der Bundesregierung* (*German Federal Government Blockchain Strategy*) and determined action measures in five major fields to accelerate the development of German blockchain industry and guide & underline the application of blockchain into noncurrency. United States ratifies blockchain technology and encourages its development; however, it bears cautious attitude and strictly supervises the emerging technology. Presently, United States firmly cracks down on illegal acts in cryptocurrency in its supervision of blockchain, yet it takes more and more positive stance toward the application of blockchain.

2. Public Infrastructure for Blockchain Becomes the Strategic Center

Major countries and regions in the world actively explore public service infrastructure for national and regional blockchains. Some EU member states signed an agreement and planned to establish European Blockchain Service Infrastructure (EBSI), a network infrastructure for distributed nodes that connects European counties, to support European public administration institutions and financial services. Its first node was launched in Belgium in February 2020. In September 2019, Sao Paulo of Brazil announced the use of blockchain technology to establish proper registration institutions for public infrastructure projects. Since the end of 2018, Estonia has provided citizens with public services on secure blockchain digital identity via Keyless Signature Infrastructure (KSI).

### 2.3.5 Domain Name Market and IPv6-Construction Advance Solidly

1. The Scale of Global Domain Name Market Continues to Grow

In general, global domain name registration market shows a favorable trend, with all kinds of Top-Level Domains (TLDs) rising year on year. According to the data of VeriSign, by March 2020, the scale of global domain name registration market reached around 367 million, with an increase of 4.5 million over the fourth quarter of 2019, by 1.2%. The number of registered domain name increased 14.9 million, by 4.2%. The scale of ccTLD, gTLD and new gTLD market all continued to grow.

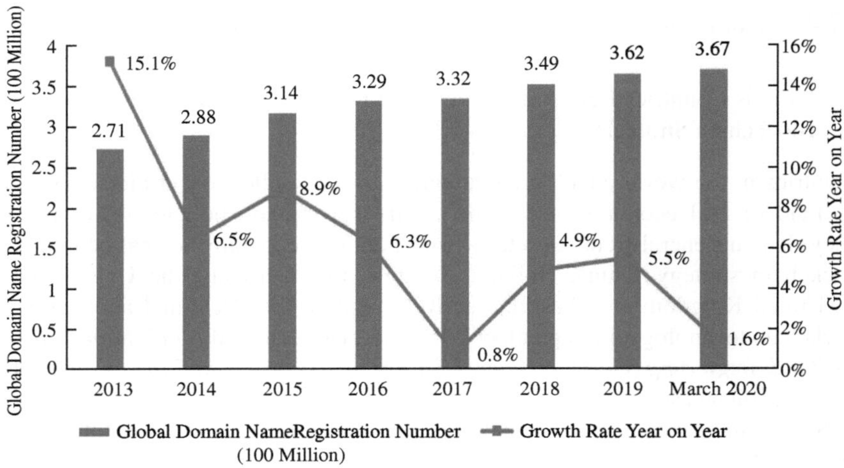


**Fig. 2.4** Statistics of global domain name registration number in 2013–2020 (March). (*Data Source* VeriSign)

Among them, ".com" and ".net" top-level domain name registration reached 160 million, with an increase of 5.9 million over the same period, by 3.8%. See Fig. 2.4 for "Statistics of Global Domain Name Registration Number in 2013–2020 (March)".

2. The Number of Root Server Nod Grows Year by Year

Root extension remains an important way to improve the efficiency and security of Internet domain name resolution, and the number of global root servers keeps growing. According to the data of Internet Corporation for Assigned Names and Numbers (ICANN), by March 2020, there were 1,253 global root servers and their mirror servers, up by 15.4% year on year, which covered over 140 countries and regions and provided global users with the nearest root resolution service. Among them, in the number of root mirror server, the United States (239), Brazil (40) and Germany (37) rank the first, the second and the third respectively. China has approved 13 root mirror servers, ranking the 19th in the world. With the rapid development of ICT business, most root server operators establish global distributed architecture by building mirror servers, which will remarkably improve the efficiency of domain name resolution and website access speed.

3. Global IPv6 Distribution Number Continues to Increase

The limited IPv4 network address resources cannot support the rapid development of telecommunications; therefore, IPv6 technology that meets the demand for massive addresses from such business as IoT and cloud computing in the future develops quickly. According to the data of Number Resource Organization (NRO), by March 2020, the total number of global IPv6 allocated address realized 304,956 (/32), up by 10.1% year on year. China and the United States have 47,860 (/32) and 54,944(/32) IPv6 allocated addresses respectively. In 2013–2019, China's IPv6 address allocation

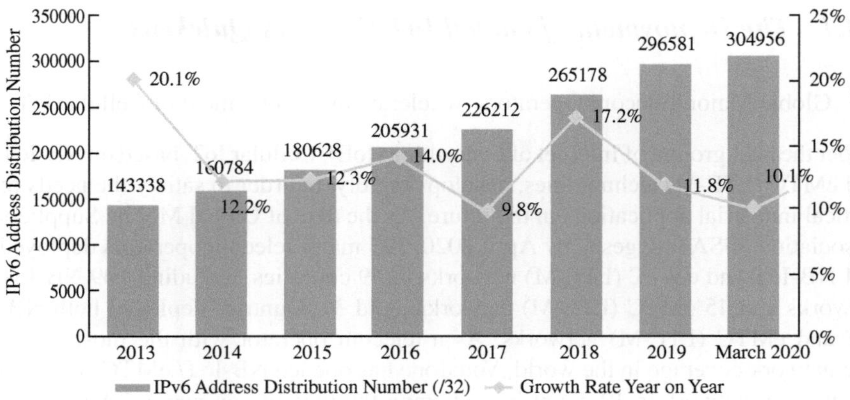


**Fig. 2.5** Global IPv6 address distribution number in 2013–2020 (March). (*Data Source* NRO)

rose significantly, with a Compound Annual Growth Rate (CAGR) of 19.2% (the United States 5.9% only). See Fig. 2.5 for "Global IPv6 Address Distribution Number in 2013–2020 (March)".

4. Global IPv6 Deployment for Commercial Use Is Actively Promoted

Global IPv6 deployment progresses fast. The deployment rate of IPv6 improves steadily in major developed countries and advances simultaneously in some developing countries. According to the data of Asia–Pacific Network Information Centre (APNIC), by May 2020, in deployment rate of IPv6, India (68.5%), Belgium (57%) and Mayotte (54%) ranked the first, the second and the third respectively in the world. China's deployment rate of IPv6 reached 16.2% and ranked the 43th in the world. Global IPv6 users are large in scale, with active users increasing significantly. As Asia–Pacific Network Information Centre (APNIC) predicted in March 2020, three countries had the greatest number of IPv6 users in 2020, India (360 million), the United States (140 million) and China (130 million).

## 2.4 The Global Deployment of New Facilities Speeds Up

The global deployment of new facilities deepens, and global major telecom operators accelerate the construction of Cellular IoT. IoT application market represented by smart city, Internet of Vehicles (IoV) and smart home remains active. The development of global industrial Internet displays a noticeable trend, faster breakthrough in key technologies, better basic support, more diverse integration application and better industrial ecology.

### 2.4.1 The Deployment of Global IoT Facilities Quickens

1. Global Major Telecom Operators Accelerate the Deployment of Cellular IoT

Under the background of Internet of Everything (IoE), Cellular IoT, based on NB-IoT and eMTC (LTE-M) technologies, develops rapidly, in order to satisfy the needs of vertical industrial applications in the future. As the data of Global Mobile Suppliers Association (GSA) suggests, by April 2020, 123 major telecom operators deployed 154 NB-IoT and eMTC (LTE-M) networks in 59 countries, including 109 NB-IoT networks and 45 eMTC (LTE-M) networks. And 31 countries deployed both NB-IoT and eMTC (LTE-M) networks. As a telecom operator with the most extensive network coverage in the world, Vodafone has opened NB-IoT/ eMTC (LTE-M) commercial network in 12 countries, allowing the roaming on nearly 20 networks globally. According to statistics of Chinese government, China's three major telecom operators have completed NB-IoT network coverage in urban areas at & above county level, deployed more than 700,000 NB-IoT base stations and developed more than 1 billion terminal users.

2. The Number of IoT Sensing Terminal Connection Increases Substantially

With the arrival of 5G era, global IoT moves fast from the initial stage (featured with fragmented and isolated application) to the stage of large-scale connection (featured with "emphasizing key application while realizing crossover integration"). As the data of Global Mobile Suppliers Association (GSA) suggests, global Cellular IoT connection grew by 43% in 2019, with the total number approaching 1.7 billion. Among them, China Mobile realized 880 million Cellular IoT connections, ranking the first in the world; Vodafone had more than 87 million Cellular IoT connections, and AT&T had more than 70 million Cellular IoT connections. As the data of Global Mobile Suppliers Association (GSA) shows, the number of global Cellular IoT connection will exceed 3.5 billion by 2025, and the Asia–Pacific Region becomes the largest IoT connection market, followed by North America and Europe. No matter in total number or growth rate, IoT has become strategic business with remarkable value for global telecom operators. See Fig. 2.6 for "The Number of Cellular IoT Connection in the World and China in 2015–2025".

### 2.4.2 The Construction of Global IoT Platform Accelerates

1. IoT Platform Gradually Becomes the Core Part of Industrial Ecology

As emerging technologies like 5G and industrial IoT develop quickly, the industrial ecology of IoT platform gradually takes shape and the competition in market increasingly intensifies. The core of IoT platform gradually turns from original Connectivity Management Platform (CMP) to Device Management Platform (DMP) and Application Enablement Platform (AEP). According to the report of Verizon, the market scale

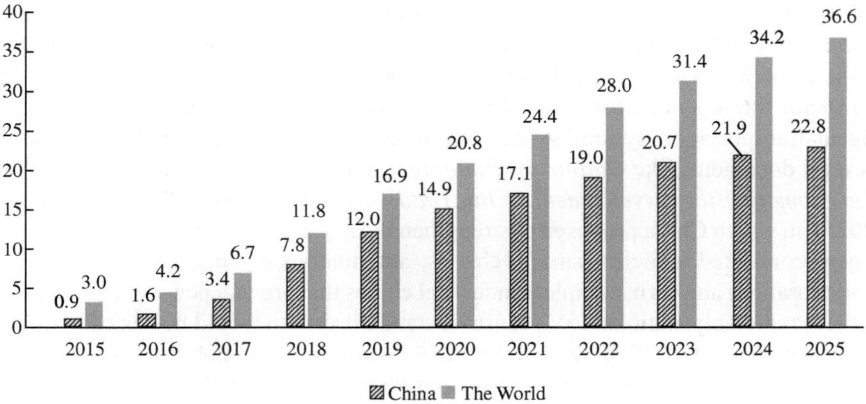


**Fig. 2.6** The Number of cellular IoT connection in the world and China in 2015–2025 (100 Million). (*Data Source* Global Mobile Suppliers Association)

of IoT platform would reach $1.6 billion by the end of 2020. Presently, Connectivity Management Platform (CMP) has been basically standardized, with communication equipment suppliers dominating platform market. Major CMPs include Jasper CC Platform, Ericsson DCP, Vodafone GDSP and Huawei Hilink Platform. Device Management Platform (DMP) develops earlier and comes into the market with Connectivity Management Platform (CMP). Yet, Application Enablement Platform (AEP) stays in initial stage and becomes the central issue in the competition among large, medium and small enterprises & startups. AWS, Microsoft, IBM Watson, and ThingWorx /PTC all concentrate on Application Enablement Platform (AEP). As IoT applications are actualized quickly, AEP will enter a stage of rapid development.

2. IoT Quickly Empowers the Construction of New-Type Smart City

Major cities in the world successively formulate new strategies that promote the level of public services through IoT technology. According to the report of Frost & Sullivan, global market value of smart city that combines IoT with AI will exceed $2 trillion by 2025, among which the market value of top 600 smart cities accounts for 60% of global GDP. With its technological advantages in chip, software, Internet and other fields, United States vigorously promotes the applications of Radio Frequency Identification (RFID), sensor, and M2M under the framework of smart city, which have been widely used in electricity, industry, agriculture, environmental monitoring, construction and medical treatment. In European Union, the use case of IoT centers on electricity, transport and logistics. Japan applies IoT early and takes Internet of Vehicles, smart town and green ICT as IoT demonstration projects.

3. Various Countries Successively Release Policies to Boost the Development of Internet of Vehicles

Internet of Vehicles (IoV) serves as an important field for the deep integration of informatization and industrialization and a key direction for the vertical application of 5G. It possesses great potential in industrial development, huge market space in application and considerable social benefit. Countries in the world release relevant policies on IoV successively. In July 2019, the United States released the *National Artificial Intelligence Strategy 2019*, which proposed to construct powerful intelligent transportation system. After September 2019, China successively released several documents like *Outline for Building a Transport Power* and *Action Outline for Promoting the Development of Big Data on Comprehensive Transport (2020–2025)*, in which China proposed to strengthen the research and development of intelligent connected vehicles (smart vehicles, autonomous driving, and vehicle–road collaboration) and form complete industrial chains that are independent and controllable. Predictably, national IoV industrial standard system would be basically established by the end of 2020. In October 2019, European Automobile Manufacturers' Association stated at World Intelligent Connected Vehicles (WICV) Conference that European Union planned to realize autonomous driving in some scenarios in 2020, Internet-connection for all vehicles by 2022 and fully automated travel by 2030.

4. Smart Home Market Displays Great Potential for Future Development

As an emerging industry, smart home stands at the transition from emergence to growth. As related intelligent applications are gradually actualized and consumer's use habits are formed, smart home market will face a stage of large-scale outbreak. According to the market research report of Mordor Intelligence, global smart home market scale will continually grow, with Compound Annual Growth Rate (CAGR) reaching 25% in 2019–2024. As the public attach more attention to home security, "security and control" devices will grow fastest in smart home applications. Globally, the United States plays a vital role in driving the consumption on global smart home market, with much more demands than other countries. As *Digital Market Outlook (DMO) Analysis Report* released by Statista estimated, in 2020, the United States would expend $24.972 billion on smart home, which accounted for 29.5% of the total expenditure in global smart home sector.

### 2.4.3 The Construction of Industrial Internet Makes Significant Progress

1. Initial Results Have Been Achieved in the Construction of Network Facilities for Industrial Internet

Industrial Internet serves as network infrastructure for constructing all-round interconnection among human, machine and material under industrial circumstance. It becomes key motive force for the high-quality development of industrial manufacturing in various countries. Global major telecom operators quicken the construction of WAN (Wide Area Network) for enterprises with low delay, high reliability and large bandwidth. Meanwhile, network technologies like 5G, edge computing, time-sensitive network, software-defined network and IPv6 constantly develop and gradually meet the high demands for production and control. Chinese telecom operators use 5G technology to transform LAN (Local Area Network) for internal Internet, which effectively promotes the wireless, de-layered and IP-based LAN development. In the construction of WAN (Wide Area Network), Chinese telecom operators construct WAN with low delay and high reliability by transforming & upgrading the existing network and building new-type network. Deutsche Telekom actively propels industrial digitization. In April 2020, Deutsche Telekom collaborated with Ericsson on launching the first 5G SA special network, which helped accelerate a series of connected network infrastructure projects on Industry 4.0.

2. Major Countries Issue Supporting Policies on Industrial Internet

In line with the reality of industrial development, major countries quickly establish and improve the industrial structure, policy framework and management system that accord with developmental trend in the future. China, Russia, South Korea and Italy successively issue policies and documents to promote the development of industrial Internet. In March 2020, Ministry of Industry and Information Technology of China released *Notice on Accelerating the Development of Industrial Internet*, which clearly presented 20 specific measures in six aspects, including accelerating the construction of new-type infrastructure, accelerating and expanding the application of integrated innovation and accelerating the improvement of security & guarantee system. In April 2020, Russian government approved *Key National Strategy for the Development of Manufacturing Industry*, which aimed to create globally competitive industrial sectors with high potential. As it conceived, the proportion of manufacturing in GDP would increase from 14 to 17%. In May 2020, South Korean government released *Intelligent Robots Action Plan 2020*, which planned to provide 1,500 advanced industrial robots for manufacturing and service industries and promote the integrated innovation of 5G, AI and robot.

3. Industrial Internet Logo Analysis System Gradually Improves

Logo Analysis System constitutes important network infrastructure of industrial Internet. It serves as a key hub to achieve the data circulation and information interaction among industrial enterprise and basic condition to realize data sharing among different systems, enterprises and regions. Industrial Internet Logo Analysis System actualizes the flexible differentiation and information management of whole-network resources, by giving the sole "ID card" to each entity object (product, part, machinery and equipment, etc.) and virtual assets (model, algorithm, process, etc.). It plays a pivotal role in promoting a new-round reform in global industrial system. Now, a variety of relatively independent Logo Analysis System has formed globally, such as Handle,[1] OID,[2] GS1[3] and Ecode.[4] Among them, Handle has been successfully applied and practiced in the fields of digital publishing and digital content. Ten Handle root systems have been set up in the world, which are respectively deployed in the United States, China, Germany, the United Kingdom, Saudi Arabia, South Africa, Rwanda, Tunisia, Russia and International Telecommunication Union (ITU). OID has been adopted by more than 200 countries and regions in the world and widely used in the fields of information security, network management and medical imaging. According to the annual report of GS1, GS1 owns 114 regional membership institutions globally, covering major countries on Five Continents. It has established mature application systems in the fields of retail, logistics and e-commerce. More than two million enterprises in the world use GS1 logo, with the average daily resolution number reaching 6 billion.

4. Industrial Internet Market Continues to Grow at High Speed

Global industrial Internet platform remains active and innovative, and starts to advance the digital transformation of manufacturing industry. As Markets and Markets predicts, the scale of global industrial Internet platform market will grow to $13.8 billion and $19.9 billion respectively in 2023 and 2025. Currently, the United States, Europe and the Asia–Pacific Region maintain their leading role in the development of industrial Internet platform. As giants like GE, Microsoft, Amazon and PTC actively deploy industrial Internet platforms and various American start-ups continuously boost the innovation on cutting-edge platform technologies, American industrial Internet platforms display prominent group advantages. After the United States come European industrial giants, such as Siemens, Schneider and Bosch, which accelerate the development of emerging industries based on their advantages in leading manufacturing foundation. In emerging economies like China and India, the demand for industrialization continually promote the development of industrial Internet platform in the Asia–Pacific Region. In addition, in the global fight against the COVID-19 Pandemic, the cloud computing and data resources of industrial Internet platform effectively met the demands for timely development&deployment

[1] Handle System refers to a global distributed management system that is organized and managed by International DONA Foundation. As the main concrete form of Digital Object Architecture (DOA), it adopts segmental management and resolution mechanism to realize the registration, resolution and management of objects.

[2] Object Identifier (OID) is an identification mechanism jointly proposed by the ISO/IEC and ITU. It is used to name objects, concepts or things of any types globally and uniformly. Once they are named, the name is permanently valid.

[3] GS1 (Global Standard 1) is an identification system established by International Article Numbering Association. It consists of three systems, i.e., coding system, carrier system and data exchange system, which can identify all kinds of data and information in the whole life cycle of object supply chain.

[4] Ecode is a Chinese national standard that was enforced on April 1, 2019. Its standard number is GB/T 36,605—2018.

of apps, the supply & demand docking and the resource allocation from governments and enterprises in the world in emergency response. This contributed to the rapid allocation and transportation of materials and the steady resumption of work and production.

# Chapter 3
# The Development of World Information Technology

## 3.1 Outline

Presently, a new-round scientific & technological revolution and industrial transformation thrive in the world, which greatly and deeply impacts on global economic development, social progress and human civilization. In 2020, scientific & technological innovation and iteration accelerated, interdisciplinary integration further deepened, and industrialization kept expanding. Facing the challenge of the COVID-19 Pandemic, various countries apply information technology to participate in the control & prevention of the epidemic and the resumption of work and production, by improving the screening of pathological indicators, quickening the production of anti-epidemic materials, providing online video services, and conducting team collaboration. These technological application and innovation constantly develop while solving various problems.

Basic technologies evolve iteratively and maintain high-speed development. High-performance computing (HPC) initially enters Exascale level, and low-power heterogeneous supercomputer platform rises rapidly. Chip technology gradually approaches its physical limit, and 3D stack technology is extensively used. Operation system presents the trend in which cloud end and terminal integrate and develop. Industrial software remains the platform-based and integrated development.

Remarkable innovation continues to emerge in cutting-edge technologies. Basic AI technological framework gradually stabilizes, and "intelligent+" progressively empowers traditional industries. Blockchain industrial standards are gradually optimized and applied into more fields. As it is claimed, quantum computing achieves "quantum supremacy", and the construction of quantum communication route accelerates. Biocomputing and storing gradually turn from theoretical construction to prototypical application, and BCI (Brain-Computer Interface) medicine displays its importance in clinical application. New technology & application quickly integrate with traditional industries, which continually promotes the rapid development of smart economy.



Chinese Academy of Cyberspace Studies, *World Internet Development Report 2020*,
https://doi.org/10.1007/978-981-16-9388-5_3

## 3.2 Basic Technologies

### *3.2.1 High-Performance Computing (HPC) Initially Achieves the Exascale Level*

International scientific & technological competition becomes increasingly fierce, and high-performance computing technology and application symbolize national comprehensive power. Major countries in the world constantly advance toward E-level (Exascale) computing capacity. In recent years, achievements are made in heterogeneous supercomputer platform and featured supercomputing application, which continuously enlarges the scope of high-performance computing.

1. High-Performance Computer Performs and Improves Stably

Currently, high-performance computer keeps developing toward E-level computing. In the 55th edition of the "Top 500" (June 2020) of International Supercomputer Performance Rating Organization,[1] Fugaku Supercomputer, developed by Fujitsu (Japan) and installed at RIKEN Computational Science Center in Kobe (Japan), ranked the first with the floating-point arithmetic performance of 415.53 PFlops (petaflops/per second). "Fugaku" boasts the first supercomputing system in the world that uses ARM architecture and takes the first place. In mixed precision computing that deals with machine learning or other AI application, Fugaku becomes the first E-level supercomputer in the world, with its tested performance of HPL-AI reaching 1.42 EFLops (exascale times/second). The floating-point computing capacity of entry-level systems in "Top 500" has risen to 1.228 petaflops/second, and the aggregation computing capacity of full-list systems has risen to 2.26 EFlops. In the number of system, China still ranks the first in "Top 500" (226), followed by the United States (114), Japan (30), France (18) and Germany (16). Four new systems are listed in top 10 supercomputers, in which Japan, the United States, China and European Union dominate. In overall performance, the United States (644 PFlops) outstrips China (565 PFlops) and takes the lead in full list. See Table 3.1 for "List of Global Top 10 Supercomputers (June 2020)".

2. Deep Learning Supercomputer Displays More "Green" Performance

As modern supercomputer computes faster, energy consumption becomes an important factor that restricts the development of supercomputer. Used to measure the power consumption and performance of supercomputer, "Green 500" List ranks the index most credibly now. In "Green 500" List, computing performance produced by 1 W-electricity power consumption is used as a basic index and assessed by High Performance Linpack (HPL), a benchmark, with the index calculated. In recent years, supercomputers that rank top in "Green 500" List mostly adopt GPU-technology-based systems. Japan and the United States take the lead in the field of energy-saving

[1] Data Source: Global Supercomputer "Top 500", June 2020. See https://www.top500.org/lists/2020/06/.

**Table 3.1** List of global top 10 supercomputers (June 2020)

| Ranking | Organization | Country | System | Cores of processor | Floating-point speed (PFlop/s) | Peak value/(PFlop/s) | Power consumption (kW) |
|---|---|---|---|---|---|---|---|
| 1 | RIKEN | Japan | Fugaku | 7 299 072 | 415.53 | 513.855 | 28 335 |
| 2 | Oak Ridge National Laboratory | The United States | Summit | 2 414 592 | 148.60 | 200.794 | 10 096 |
| 3 | Lawrence Livermore National Laboratory | The United States | Sierra | 1 572 480 | 94.64 | 125.712 | 7 438 |
| 4 | National Supercomputing Center in Wuxi | China | Sunway • TaihuLight | 10 649 600 | 93.01 | 125.436 | 15 371 |
| 5 | National Supercomputing Center in Guangzhou | China | Tianhe-2 A | 4 981 760 | 61.44 | 100.679 | 18 482 |
| 6 | Eni SpA | Italy | HPC5 | 669 760 | 35.45 | 51.721 | 2 252 |
| 7 | NVIDIA Corporation | The United States | Selene | 272 800 | 27.58 | 34.569 | 1 340 |
| 8 | Texas Advanced Computing Center | The United States | Frontera | 448 448 | 23.52 | 38.746 | – |
| 9 | CINECA | Italy | Marconi -100 | 347 776 | 21.64 | 29.354 | 1 980 |
| 10 | Swiss National Supercomputing Centre | Switzerland | Piz Daint | 387 872 | 21.23 | 27.154 | 2 384 |

supercomputing system and technology. In the latest "Green 500" List released in June 2020, MN-3, a supercomputer for deep learning co-developed by Kobe University and RIKEN in Japan, ranked the first. MN-3 achieves the processing performance of 21.11 GFlops/W (21.1 billion floating-point computing/per watt/per second), 1.15 times of the best performance on this list in June 2018 (18.404 GFlops/W). Systems listed in "Green 500" vie in terms of "ultra-low power consumption of supercomputers for deep learning", and the technologies of Mn-Core and Mn-3 keep ahead globally. Main technologies that can be used to achieve high power-saving performance include "MN-Core" (a special circuit for deep learning with ultra-low power consumption), "MN-Core Direct Connect" (high speed & efficiency data transmission and interconnection among nodes), technology optimizing the efficiency of deep learning-oriented double matrix multiplication computing, and technology maximizing power efficiency with massively integrated MN-Core.

3. Heterogeneous Supercomputer Platforms Integrate and Develop Quickly

Heterogeneous system architecture of modern supercomputer significantly improves computing performance with limited power consumption. Yet, it directly causes more complex problems on programming and limits the efficient use of computing system by applied scientists. Therefore, various heterogeneous programming models constantly appear. The popular ones are as follows:

(1) Compute Unified Device Architecture (CUDA), proposed by Nvidia, a developer of GPU, only supports Nvidia's GPU.
(2) Open Computing Language (OpenCL), an open standard for the parallel programming of heterogeneous systems, can undertake parallel programming for various accelerator components among multi-platforms.
(3) Portable Heterogeneous Computing Interface HIP, a vendor-neutral C++ programming model, can finish highly optimized GPU workload.
(4) OpenAPI unified and simplified heterogeneous programming is proposed by Intel. It attempts to cover extensive computing architectures like Scalar, Vector, Matrix and Spatial, which are applied into CPU, GPU, AI accelerator and FPGA component respectively.

With the development of heterogeneous supercomputing platforms, HPC (high-performance computing) and AI integrate and develop quickly. HPC+AI computing can be done with analogous software that is compatible with multi-hardware by programming with RAM interaction model and configuring HPC, AI and big data analysis with identical architecture.

4. Featured Application Enhances the Development of Supercomputer

In 2019, ACM Gordon Bell Award, a weather vane for high performance computing, was presented to the research team at ETH Zurich, which developed mappable transistor thermal simulation. By re-organizing data-centered applications, the computing simulation improves computing efficiency by at least two orders of magnitude. The codes that the application team creates achieve the sustainable performances of 85.45 PFlops (double precision) and 90.89 PFlops (mixed precision) on

4,560 nodes of Summit supercomputer. The computing simulation enables the reconstructed quantum transport simulator to process the real nano-electronic component that consists of 10,000 atoms. Compared with the former system that consists of 1,000 atoms, the computing efficiency increases by 14 times, with invariant precision and CPU/GPU number.

### *3.2.2 Chip Technology Faces Technological Revolution*

Chip technology forms one core competitiveness of modern science and technology. In recent years, as the critical size of chip gradually approaches its physical limit, Moore's Law shows a slowing trend. Influenced by such factors as the breakthrough in emerging technology, the change of market pattern and the adjustment of industrial organization, inflection point will appear in the development of chip.

1. Computing Chip Industry Clusters Together, with Chip Performance Improving Steadily

In CPU chip, global market scale realized $57.48 billion in 2019, up by 2.8% year on year.[2] Intel's X86-architecture CPU monopolizes the market of desktop computer and supercomputer & server. Intel planned to launch 11G Tiger Lake processors in September 2020, which would replace 10 nm + process Ice Lake processor. In 2019, Zen2-architecture Ryzen processor that AMD manufactured based on tsmc's 7 nm process improved its product performance and enlarged its market share.

In GPU chip, global market scale reached $7.434 billion in 2019, down by 10.3% year on year.[3] With such advantages as short cycle in design and development, strong parallel computing capacity and complete software ecology, GPU serves as mainstream international cloud-end technological roadmap. Nvidia and AMD basically monopolize the global GPU market. Nvidia monopolizes the market of chip that accelerates the algorithm of deep learning and applies the strong HashRate performance of GPU into many fields like AI, intelligent driving and high-performance platform. In December 2019, Nvidia launched the latest version of TensorRT 7 at GPU Technology Conference. In May 2020, Nvidia launched Tesla A100 chip that depended on brand-new Ampere architecture and tsmc's 7 nm process and adopted 3D stacking technology. With chip performance remarkably improved, it accounted for 18.3% of the market share. Three manufacturers, Qualcomm, ARM and Imagination dominate the mobile GPU market, and Qualcomm's GPU has been integrated into Snapdragon chip.

In DSP chip, global market scale totaled $1.28 billion in 2019, significantly down year on year. Presently, three corporations, Texas Instruments (39.6%), NXP (25.3%)

[2] Data Source: Statistical report published by Gartner: *Marker Share: Semiconductors by End Market, Worldwide, 2019.*

[3] Data Source: Statistical report published by Gartner: *Marker Share: Semiconductors by End Market, Worldwide, 2019.*

and ADI (22.4%) monopolize global market. Hopefully, DSP chip will further improve core structure, integrate with microprocessor/high-end CPU/SoC, and deepen chip-core integration, programmability and fixed point in future development.

2. Memory Chip Market Fluctuates Occasionally and Technology Promotes 3D Stacking

Now, the size of memory technology gradually approaches the physical limit and slows down entirely owing to the problems like power consumption and heat radiation. DRAM and NAND Flash mainly constitute the memory chip industry and account for 96% of the market scale of entire memory industry. Top 5 memory chip corporations, Samsung (38.9%), SK Hynix (19.7%), Micron (18.5%), Kioxia[4] (7.1%) and Western Digital (5.6%) take up around 90% of total market share. In 2019, DRAM's market scale reached $62.17 billion, down by 37.8% year on year; NAND's market scale reached $42.63 billion, down by 26.4% year on year. The downturn in the memory market plays a major part in the decline of global semiconductor market.

In DRAM, the price fluctuates evidently, and the entire market undergoes great volatility of 30% all the year round. Now, major manufacturers develop into 1Znm technological node and display differences in the mass production. In the second half of 2019, Micron, SK Hynix and Samsung started to massively produce 1Znm-process DRAM. In high-end product, Samsung still takes the lead. In July 2020, it announced the mass production of the first 12 Gb LPDDR5 chip in the world.

After NAND Flash develops into 3D stacking, its upgrading quickens and its capacity grows rapidly. In the market, the supply–demand relationship fluctuates obviously, which often causes the great volatility of the total market share. Corporations on NAND Flash and DRAM highly overlap. As technology and process grow to 14 nm, technological difficulty increases exponentially, with 3D NAND technology as a new developmental trend. In 2019, major international corporations commonly started the mass production of 96–128 layer NAND. For instance, in June 2020, SK Hynix released the 128-layer 4D TLC NAND; in August 2020, Samsung started to massively produced 6G 128-layer TLC 3D NAND; in August 2020, Kioxia launched XL-Flash storage-level memory based on the original 4G 96-layer BiCS Flash.

3. 5G Technology Spurs the Fast Development of Communication Chip

Communication chip mainly include baseband chip and radio frequency chip. As 5G technology stimulates the development of integrated circuit industry, communication chip faces new development opportunity. According to the forecasting data of Gartner, global penetration rate of 5G smartphone will rise from 1% in 2019 to 60% in 2023, which triggers the rapid growth of baseband/application processor, RF front end, power management chips and other products. Major semiconductor corporations all deploy in 5G field. Qualcomm, Huawei, MediaTek and Samsung have all released 5G baseband and SoC chip.

[4] It was formerly called Toshiba Memory Corporation, which was officially renamed Kioxia in October 2019.

In baseband chip, the solution that integrates baseband and application processor into SoC chip presents a mainstream developmental trend. Representative products are Qualcomm's X50/X55/X60, Samsung's Exynos M5100/M5123, MediaTek's Helio M70, Huawei's Balong 5000, Huawei's Kirin 990 5G and Unisoc's Chunteng 510. In global market, Qualcomm, Huawei Hisilicon, Intel, MediaTek and Samsung ranked top 5 in baseband chip in 2019. With second-generation 5G products like X55 ultra-thin modem and Snapdragon 765/G 5G SoC, Qualcomm ranked the first and realized an income share of 41%. The COVID-19 Pandemic and sagging seasonal demand impaired baseband shipment and the shipment of 4G baseband chip decreased to some degree; however, segmented market still gave opportunities to baseband chip suppliers, and the high price of 5G baseband chip continually promoted the growth of income in the baseband ship market.

In RF chip, filter and power amplifier (PA) are two largest segments in the field of RF front-end, which account for 61% of the RF front-end market. Presently, the total market scale of global RF front-end increases steadily, with high concentration ratio. International giants like Qrovo, Skyworks and Broadcom mainly monopolize the market. Besides, global RF corporations continue to expand, integrate and develop channels for technology, product and market. RF chip is widely used in compound semiconductor. In its development, SAW filter is replaced by BAW filter, and RF chip is integrated.

4. Innovation on Chip Process Concentrates on New Structure & Material

The integration level of integrated circuit continuously improves, and technology & process approaches the physical limit. As the key size continues to narrow, new device structure (e.g. FinFET probably replaced by GAA) and lithographical device (from DUV to EUV) & material (probably Co, a metal, as the core of future IC metal layer) drive the development of process nodes toward smaller size. Meanwhile, as the capital invested into advanced manufacturing production line substantially jumps, the increase of process cost causes the dramatic drop in the number of advanced process enterprises. UMC and Global Foundries have successively announced to abandon the research and development of 10 nm and below-10 nm processes. In the future, in the field of advanced process-technology development, there will be only three leading corporations, tsmc, Samsung and Intel.

Considering the proportion of OEMs' business incomes from various key sizes, global OEMs realized the sales revenue of $56.5 billion in 2019, in which sizes under 40 nm took up the large proportion, rising from 40% in 2018 to 47% in 2019. The proportion of contract manufacturing with sizes of 16/14 nm and below keeps rising. The sales revenue of 16 nm and below advanced technology & processes accounts for as high as 50% of tsmc, a leading corporation in contract manufacturing. Specifically, the market share of 7 nm-process chip soared after it was massively produced in the third quarter of 2018. In the fourth quarter of 2019, its market share reached 35%.

In manufacturing technology, the advanced technology process continues Moore's Law. International leading manufacturing corporations like Samsung and tsmc maintain the leading position in advanced technology process and advance toward 5 nm,

3 nm and even 1 nm process nodes. As early as 2020, tsmc achieved the mass production of 5 nm process and firstly starts the risk mass production of 3 nm process chip in 2021, as predicted. It makes great breakthrough in 2 nm-process R&D by introducing Gate-All-Around (GAA) technology. In April 2019, Samsung announced the completion of 5 nm-technology R&D, which would be massively produced in the end of 2020, as it predicted.

Technologically, as the linewidth of semiconductor device process has been reduced to less than 20 nm, planar MOS transistor made by traditional technology meets various challenges, such as lithographical technology, high-k insulation-layer technology and power consumption. Fin Field-Effect Transistor (FinFET) developed by Intel and Fully Depleted Silicon on Insulator (FD-SOI) developed by IBM are widely used in technological route. Currently, FinFET technology realizes the mass production of 7 nm process and the risk production of 5 nm process (the mass production in 2020). FD-SOI technology that has obvious advantages in high frequency, low power consumption and anti-static electricity recently makes a breakthrough in 12 nm process.

### *3.2.3 The Integration and Intelligentization of Software Technology Develop Faster*

Presently, as Moore's Law slows down, software promotes the development of information technology more strongly. The integration and *intelligentization* of cloud (server cluster), side (intelligent terminal) and end (IoT equipment) display a remarkable trend, in which operating system plays a core role. The new-round competition between programming language and compiling toolchain takes shape, which remaps the application market. The strategic role of industrial software is highlighted, with open source as the main model of ecological competition.

1. The Industrial Pattern of Operating System Stabilizes and Continues to Develop Diversely

As the foundation of software pattern, operating systems mainly covers server/cloud operating system, terminal operating system and IoT operating system. With the development and interaction of cloud, terminal and IoT devices, operating systems present the developmental trend of diversity in form and integration in technology.

(1) Industrial giants start industrial integration in server operating system/cloud operating system. Now, enterprise-edition RHEL, open-source edition CentOS, and community-edition Fedora of Red Hat (United States) take the lead. In July 2019, IBM officially announced it purchased Red Hat with around $34 billion. Canonical (the United States) constantly concentrates on Ubuntu operating system. Huawei makes Euler operating system as open source and builds openEuler open source community. In the field of cloud platform, container technology represented by Docker becomes more popular, and Kubernetes, an

open source container orchestration engine, becomes steadier. Serverless architecture (event-driven cloud function, no need of server manager) that developes from Microservice remains a hot topic. Serverless architectures dominated by Amazon AWS and Microsoft Azure compete fiercely.

(2) Terminal operating system mainly includes mobile end and desktop end, both of which make new progress. In the field of mobile end operating system, Android and iOS rule the roost and monopolize firmly, and other operating systems take up less than 1%.[5] The source code of Fuchsia, a new Google operating system, partially serves as open source. Fuchsia uses Zircon, a microkernel different from Linux, and works with DART language and Flutter mobile application development framework. Hopefully, it runs on multiple intelligent terminals across platforms and better supports VR/MR in the future. In the field of desktop operating system, Windows accounts for 77.04% in market share and macOS 18.38%.[6] To make Windows appeal to developers, Microsoft actively introduces WSL (Windows Subsystem for Linux), which supports the running of original Linux (distribution edition) on Windows via virtualized technology. ChromeOS, which Google publicizes vigorously, grows a slower speed than expected. Other distribution editions of Linux are optimized according to their application fields. Wayland rendering engine is widely used in Graphical User Interface (GUI), yet this advances slowly. No matter terminal or desktop, the inherent ecology of operating system shows the noticeable hesitation effect before new technology, as ecological monopoly remains hard to change.

(3) The ecological integration of IoT operating system accelerates. As concepts like IoT and edge computing become popular and terminal computing capacity improves, more and more computing tasks are performed on terminal devices, and IoT operating system plays a decisive role in the ecology. Currently, major IoT operating systems abroad include FreeRTOS (developed by Amazon) and MbedOS (developed by ARM). In China, IoT operating systems include LiteOS (Huawei), AliOS Things (Alibaba) and TencentOS-Tiny (Tencent). As IoT operating system aims to achieve light weight and low power consumption, it looks simple in structure yet involves a wide range of hardware. It is generally tied up with cloud platform manufacturers to extend their product and service. Besides, RISC-V instruction set comes to rise and starts from IoT, which creates a new opportunity for ecological development.

(4) Opportunities for the reform in multi-platform integrated operating system takes shape. Microsoft completes the overall layout in Windows operating system, from server, desktop to intelligent terminal. Though Microsoft fails in the field of intelligent terminal, it achieves high-degree reuse of server and desktop operating systems through Azure cloud computing platform. Apple launches iPados, a new operating system independent of macOS and iOS,

---

[5] Data Source: Statcounter. Mobile Operating System Market Share Worldwide-August 2020[ol]. https://gs.statcounter.com/os-market-share/mobile/worldwide.

[6] Data Source: Statcounter. Desktop Operating System Market Share Worldwide-August 2020[ol]. https://gs.statcounter.com/os-market-share/desktop/worldwide.

whose in-laid Sidecar function connects applications across various hardware platforms. Now, the demands for cross-scene, cross-physical region, cross-device and cross-architecture open new opportunities for technological reform in operating system. Integrated iterative development and intelligent deployment & operation become the internal demands of operating system manufacturers, which will speed up the internal integration of operating system technologies.

2. New-Round Competition Takes Shape between Programming Language and Compilers

Programming language & compiler, running environment and integrated development environment determine the development model, method and tool of high-level application. They are foundation for establishing dominant position in application software ecology, for which major giants vie in addition to operating system.

(1) There are various programming languages with unique features. Presently, C/C++ language remains the first choice for system software, Python language dominates AI field, and Java/JavaScript takes the lead in the Internet and mobile Internet application. In recent years, Google has striven to develop the server program with GO language, to replace Java with Kotlin as the preferential language for App development on Android operating system, to use Chrome V8 engine as the dominant engine for JavaScript (based on which node.js as preferential runtime framework for Web platform), and to use DART as default application programming language for Fuchsia, a new-generation operating system. Swift programming language that Apple launched ranked top 10 in Popular Programming Languages List 2020.
(2) New and old compilers co-exist and AI-related optimization becomes a hot topic. GCC compiler, which came to rise in the 1980s, looks massive and dominates GNU/Linux systems now. LLVM compiler started to be used in 2000. With the strong support from Apple and Google, it stands comparison with GCC compiler. In recent years, with the popularization of AI computing represented by neural network, compiler played an important role in unifying algorithmic model and operator representation and executing the HashRate of AI-dedicated chip. In 2019, Google made MLIR, a brand-new media code & compiler framework, as open source, which can be adapted to a variety of hardware, like GPU, TPU and mobile devices.
(3) The competition on Integrated Development Environment (IDE) intensifies, and open source becomes popular. Eclipse open source editor that IBM developed early continues to iterate. Android Studio that Google develops remains the first choice for App development on Android operating system. Now, IntelliJ IDEA (JetBrain) boasts the most popular Java development environment. On the basis of Visual Studio, Microsoft launches and dominates VS Code open source project, which becomes the most active open source integrated development environment project on GitHub now. It supports the development of multiple applications like IoT, mobile terminal and server

by providing various plug-ins. Opened as open source and supporting third-party plug-in become the core competitive model of integrated development environment products.

3. Industrial Software Plays a More Prominent Role with Noticeable Industrial Monopoly

Industrial software serves as a productive tool in the field of industrial segmentation. It is important to improve the productive efficiency of industrial software. The representative ones include EDA (Electronic Design Automation) software for chip design, tool software for numerical computing and simulation modeling, and CAD/CAE software for user computer-aided design and manufacturing.

(1) Several corporations monopolize EDA field. EDA software used in chip design is a key chain of chip industry. Globally, Cadence, Mentor and Synopsys account for more than 70% of market share. With the rise of chip process, the complexity and technological threshold of EDA software increase, which strangles the progress of latecomers.
(2) AI and open source create new opportunities for the reform in EDA field. AI technology starts to be used in EDA design. For example, Google uses AI algorithm to learn the planning and layout of chip and designs products better and quicker than manual design. Open source EDA software also develops fast. In addition to the existing gEDA, UC Berkeley launches Chisel tool for RISC-V open source instruction set, which works with Firesim simulation software and probably breaks the monopoly of commercial EDA. Google also collaborates with Skywater on launching PDK, a toolkit for open source chip process design.
(3) Numerical computing & simulation modeling software presents a three-sided pattern. MATLAB, Mathematica and Maple are collectively called "three major mathematical modeling software in the world". GNU Octave is an open source software under General Public License (GPL) that takes MATLAB as benchmark and excludes Simulink component. SCILAB is an open source scientific & engineering computing software developed by French scientists. There is certain gap between SCILAB and MATLAB in function and performance.
(4) Computer-aided software CAD/CAE displays remarkable Matthew Effect. AutoCAD from AutoDesk (the United States) is the most popular automatic computer-aided design software. Its Fusion 360 realizes whole-process industrial design assembly line and maintains competitive advantage. Currently, ANSYS (the United States) monopolizes the field of international finite-element analysis, and ANSYS 2020 R2 that it newly launches enhances team collaboration again.
(5) AR/VR become new growth point in CAD/CAE software. AR/ VR-based game engine software like Unreal and Unity gradually expands toward industrial field. For instance, in May 2020, Unity bought Finger Food, a Canadian corporation on industrial application software.

## 3.3 Cutting-Edge Technologies

### *3.3.1 AI Keeps High-Speed Development*

In recent years, stimulated by three factors, i.e. data, algorithm and HashRate, AI entered a stage of rapid development. As Internet Date Center (IDC), an international research agency, predicted, global AI market scale would reach $156 billion in 2020, up by 12.3% year on year. In specific, as the largest business, AI software accounted for 80%. Globally, leading countries China and United States form the first tier in AI, and developed countries like Japan, the United Kingdom, Israel and France form the second tier.

1. Various Countries and International Organizations Issue AI Strategies

Major countries in the world gradually turn from the spontaneous and decentral model of free exploration into the industrialization & application-oriented innovation model driven and led by national strategies. In June 2019, G20 Ministerial Meeting passed *G20 Principles on AI* to promote the establishment of national policies and international cooperation on reliable AI. In June 2019, United States published *National Artificial Intelligence Research and Development Strategic Plan* (Updated Edition), which updated original seven strategies to eight strategic priorities for investment and R&D and laid stress on in-advance layout of algorithm and chip. In February 2020, European Union announced *Data and Artificial Intelligence Strategy*, which proposed the developmental goal of openness, justice, diversity, democracy and confidence, and clarified the idea and action for digital reform in the future, in order to ensure that digital reform benefits all.

2. AI Algorithm Continuously Makes Breakthrough

The efficiency of AI algorithm continues to rise at high speed. According to the report released by OpenAI, the computing load by which neural network can be trained to fulfill identical performance in ImageNet classification decreases by 50% every 16 months. Compared with 2012, the computing load by which neural network can be trained to meet the needs of AlexNet decreases to 1/44. For general AI computing tasks, algorithmic progress produces better results than hardware performance.

(1) In computer vision, large-scale unsupervised and self-supervised models become popular, and to reduce the amount of data for learning and utilize unlabeled data become a hot research topic. The research team from the University of Oxford develops single-image method that can identify symmetrical-structure-based 3D deformable objects,[7] and accurately restore 3D shapes of human face, cat's face and vehicle in monocular images without any supervision or prior shape model. The teams from Simon Fraser University and Google Research develop BSPNet-based unsupervised learning method,[8]

[7] Wu et al. (2020).

[8] Chen et al. (2020).

which obtains a set of convex reconstructed shapes by using BSPtree based on a set of 2D. Compact in shape (i.e. low polygon), the generated grid perfectly represents sharp geometries. A collaborative team of German scholars and Facebook proposes multiview-supervision-based deep learning method for weak supervision human-body expressing & capturing.[9] It does not use 3D Ground Truth label, as two independent networks on which network architectures are based resolve human body into posture estimation and non-rigid surface deformation, so as to express and capture human body in the best way at the moment.

Neural Rendering technology that combines graphics with deep learning becomes a new model with great potential. Data-driven graphic system can synthesize new image by using multiple-image region and extract typical semantic feature based on large-scale image data set. Facebook develops DeepFovea for AI gaze point rendering, which significantly improves rendering efficiency for All-in-One VR.

The theoretical analysis of domain adaptation, Generative Adversarial Networks (GAN), Graph Neural Networks (GNN) and other technologies gradually deepens. The in-depth analysis of such theoretical problems as the mode collapse and convergence of GAN, the explanation to the principle of GNN algorithm, variant model, the expansion and adaptation of various graphic data and other tasks successively come to rise.

(2) In natural language processing, the number of language model parameter reaches new highs repeatedly. In August 2019, Nvidia released MegatronLM, a language model with 8.3 billion parameters. In February 2020, Microsoft released Turing-NLG with 17 billion parameters. In July 2020, OpenAI in Silicon Valley released the latest GPT-3 model with 175 billion parameters. So far, these models achieve the best effect in some computing tasks, yet their colossal data and computing load become a barrier for small development teams to enter the field.

The pre-training and fine-tuning methods like BERT and GPT become popular. For example, the team from Google designs a lightweight Bert[10] model that effectively solves the problem of model communication overhead, by compressing the amount of model data. A research team from The University of Hong Kong and Huawei proposes Perturbed Masking, a completely unsupervised syntactic analysis method based on the pre-trained BERT (Bidirectionl Encoder Representation from Transformers),[11] which can quantitatively analyze to what degree AI trains syntactic structure.

Interpretable natural language processing (NLP) receives more and more attention. The team from Microsoft Research raises CheckList, a method for task-independent NLP model testing, which includes some common language capabilities and test types that promote comprehensive testing. Besides, it

[9] Habermann et al. (2020).

[10] Lan et al. (2019).

[11] Wu et al. (2020).

includes a software tool that can quickly generate many different test cases, from which lots of hidden Bugs can be discovered in traditional models.

Knowledge graph becomes more popular and retrieval returns. Scientific researchers from UIUC (University of Illinois at Urbana-Champaign, US) and other institutions develop the first comprehensive open source multimedia knowledge extraction system, which can create the coherent and structured knowledge base, index entity, relationship and event, by extracting massive unstructured heterogeneous multimedia data and following the rich fine-grained entity, based on various content sources and languages. The research proposes Gaia, an operating system that can achieve the seamless search of complex graph Query and retrieve multimedia information like text, image and video.

3. Industrial Giants Monopolize Medium and High-End AI Chip

Major technological routes of AI chip include general GPU, half -customized FPGA and fully-customized ASIC. Different types of chips possess distinctive advantages and present the trend of parallel development of multiple technological routes in different fields. Basic layer of chip features with great difficulty in innovation and high barrier in technology and capital; therefore, underlying basic technology and high-end product market are monopolized by several international giants. The design and production of GPU mature and become the first choice for AI chip. In 2019, Nvidia launched TeslaV100 and TeslaT4, which displayed extremely high performance and strong competitiveness and consolidated Nvidia's monopolistic status. Duopoly continues in FPGA market: Xilinx and Intel, whose market share approximate 90%.

As a customized chip designed for the needs of special users, ASIC can be used in various terminals. At early stage, ASIC requires lots of physical design, time, capital and verification, and often displays advantages in performance, energy consumption, cost and reliability after mass production. Presently, the competition pattern of ASIC chip market remains stable and decentralized, typically represented by Google TPU. Additionally, new customized AI chips in 2019 included IPU developed by Graphcore (United Kingdom), Inferentia developed by Amazon and Nervana NNP developed by Intel.

4. Open Source Deep Learning Framework Gradually Develops in a Converged Way

PyTorch and TensorFlow, major open source deep learning frameworks, gradually enlarge their gap with competitors and converge. In October 2019, TensorFlow 2.0 was formally released, which mainly improved GPU acceleration, automatic derivation and neural network application program interface (API). In April 2020, PyTorch 1.5 was formally released, which mainly improved more flexible front-end API integration.

(1) PyTorch plays an increasingly dominant role in the field of scientific & academic researches. In 2019, PyTorch took up the overwhelming proportion. According to the statistics,[12] PyTorch was used in 69% of IEEE CVPR

[12] Horace He, The State of Machine Learning Frameworks in 2019[OL], The Gradient, 2019.

(Computer Vision and Pattern Recognition) conference papers, over 75% of NAACL and ACL conference papers and over 50% of ICLR and ICML conference papers. PyTorch was frequently used in conferences on computer vision and language (outstripping TensorFlow with the ratios of 2:1 and 3:1 respectively), and was more popular than TensorFlow at general machine learning conferences like ICLR and ICML. With the exception of ICML conference, the growth rate of papers that use TensorFlow in academic conferences cannot overtake that of the overall paper number.

(2) TensorFlow remains the leading framework in industrial circle. As the 2018–2019 statistical data suggested,[13] there were 1,541 new TensorFlow-related jobs and 1,437 new PyTorch-related jobs. In *Medium*, a famous scientific & technological magazine, there were 3,230 new articles on TensorFlow, and only 1,200 new articles on PyTorch. At GitHub, the project programmed with TensorFlow received 13,700 stars, yet the project programmed with PyTorch received only 7,200 stars. Remarkably, TensorFlow has greater advantages in productive environment.

### 3.3.2 Key Innovative Achievements Continuously Emerge in Quantum Information

Since 2019, rapid development and progress have been realized in the field of quantum information, with a number of representative achievements continuously emerging. Propelled by international technological corporations like IBM and Google, quantum computer keeps improving in bit number and manipulation precision, and gradually advances toward commercial use. In quantum cloud platform service combined with the research and development of quantum computer hardware, the development of quantum software comes from laboratory to market led by Microsoft and Amazon. Various governments and research institutions in the world continue to strengthen financial investment and policy support in the field of quantum information. Now, new pattern takes shape, in which many countries participate, theoretical and experimental researches parallel, and software and hardware develop in a coordinated way.

1. Quantum Supremacy is Actualized in Quantum Computing

In October 2019, Google AI Quantum Team announced that quantum supremacy (also called quantum advantage) was achieved on Sycamore chip, the latest superconducting quantum computing processor that it developed. Using the 53-bit quantum computer, Google completed random quantum circuit sampling experiment in 3 min and 20 s and realized 20-depth double-bit gate quantum circuit, with 430 double-quantum bit gates and 1,113 single-quantum bit gates as well as the overall fidelity of 0.2% (estimated). Google claimed that to perform the same task on Summit, the

[13] Data Source: Towards Data Science. Deep learning framework power score [OL]. https://towardsdatascience.com/deep-learning-framework-power-scores-2018-23607ddf297a.

most powerful supercomputer in the world, would take 10,000 years to compute, but IBM viewed that simulating Google quantum computer on the traditional computer would only take 2.5 days. Although the comparison of the complexity on classical simulation and the practical value of algorithm remain controversial, the research findings of Google AI Quantum Team represent a milestone in the development of quantum computer.

In 2019, major technological corporations in the world concentrated on developing and researching a series of feasible quantum computer solutions and gradually put them into commercial use. On September 18, 2019, IBM announced to establish new Quantum data centers and launch new-type 53-Qubit quantum machine. With its quantum volume untested, researchers from the University of Chicago (the United States) used the above quantum computer to create quantum entanglement in April 2020. In January 2020, IBM released the latest 28-Qubit quantum computer Raleigh, whose quantum volume[14] reached 32. Compared with IBM Q System One, the first commercial quantum computing prototype in the world that was released in early 2019, it doubled the quantum volume. In December 2019, Intel launched low-temperature control chip designated as "Horse Ridge", which achieved the control over multiple quantum bits and provided solutions for large-scale system construction. In June 2020, Honeywell started quantum computer with a quantum volume of 64, which owned the highest quality and lowest error rate of available quantum bits.

2. The Integration of Quantum Cloud Service and Hardware Develops Fast

Quantum computer poses high technological requirement and R&D cost, which impedes the popularization and marketing of quantum computer. Under such circumstance, major international corporations successively take part in the development of quantum programming language and deploy quantum cloud service. This lowers the threshold for the use of quantum computer and enlarges its use in practice. In November 2019, Microsoft released Azure Quantum quantum services, which covered quantum solution, quantum software and quantum hardware, and formed a complete open cloud ecosystem. With Azure Quantum, one got access to IonQ (from the University of Maryland, developed by Honeywell, a quantum hardware supplier) and QCI (from Yale University, a quantum computer prototype) via cloud computing platform. In December 2019, Amazon announced to launch quantum computing service Braket, and collaborated with three quantum computing corporations, i.e. D-Wave, IonQ and Rigetti. This enabled enterprises and developers to access related systems through cloud-end service, including 32-Qubit quantum computing machine developed by Rigetti. In 2020, D-Wave launched Leap 2, a hybrid quantum computing cloud platform, which was used by scholars who studied the virus of COVID-19 for free.

Developer-oriented quantum programming language that works with quantum cloud service also makes new achievements, which provides quantum computer users

[14] Quantum Volume is an index proposed by IBM and used to measure the capacity of quantum computer. Specifically, it refers to the amount of effective quantum computing that can be done in given space and time.

with standardized or visualized operating language. In March 2020, Google released quantum-edition TensorFlow that enabled developers to use classical computing technology and quantum computer route simultaneously to simulate & create hybrid AI algorithm. In April 2020, Europe started Quantum Inspire, the first public quantum computing platform, which facilitated the access to the first quantum processor in the world that used scalable spin quantum bit.

Presently, the software and hardware of quantum computing develop in a collaborative way. In addition to the development of quantum programming language and cloud service stated-above, many new achievements continuously emerge in the research on underlying hardware. In December 2019, Intel launched low-temperature control chip designated as "Horse Ridge", which achieved the control over multiple quantum bits and provided solutions for large-scale system construction to solve the extensibility of quantum bit. Meanwhile, Russian and Swedish scientists cooperated and developed a method to create quantum bit that could worked stably at room temperature and opened new outlooks for quantum computer. In April 2020, researchers at University of New South Wales (Sydney) lowered the running temperature of quantum computing platform to 1.5 K (Kelvin), which significantly reduced the cost.

3. Quantum Communication is Gradually Actualized

With the rapid development of quantum computing, quantum communication technology, an important part of quantum information, attracts great global attention, featured with absolute security, large capacity and high efficiency. Recently, in experimental and theoretical researches, a series of significant research achievements were made in the world. In October 2019, scientists from Harvard University (the United States) used quantum bits in diamond to demonstrate single-photon storage with a coherence time of more than 1 ms, which could be used as register with high-fidelity storage capacity in quantum network and effective interface for photons transmitted in optical fiber. This established foundation for the realization of quantum repeater. In March 2020, researchers from University of Oxford (the United Kingdom) reported how two remote quantum bits produced high-speed high-precision entanglement in Ion trap system, which provided new method for various applications of quantum network in the future.

In policy on quantum communication, various countries successively propose developmental plans and strategic layouts. In February 2020, National Quantum Coordination Office at the White House issued *A Strategic Vision for America's Quantum Networks*, in which it proposed that the United States would build quantum Internet to ensure quantum information science benefited the public, and planned to demonstrate basic science and key technology of quantum network in next five years, from quantum interconnection, quantum repeater, quantum memory to high-throughput quantum channel and space-based entanglement distribution that explores intercontinental distance. In addition to the United States, Japan as well as European countries and regions strived to promote the industrialization of quantum communication and deploy quantum communication in an all-round way. In January 2020, Toshiba and Tohoku University (Japan) announced that their researchers successfully transmitted complete data on human genome in a short time by using quantum private communication technology. Russia planned to use railway infrastructure to build quantum Internet platform, and started to construct pilot zone in 2021. Besides, the United Kingdom and Singapore co-established CubeSat-based quantum-encrypted satellite link, which would be put into operation by the end of 2021 as planned.

### *3.3.3 Biocomputing and Biostoring Achieve Key Breakthroughs*

Currently, major technologies of biocomputing and biostoring are DNA computing that uses natural parallel operation and biochemical processing of massive DNA molecule and DNA storage that uses four-basic group A/T/C/G permutation encoding performance of DNA molecule respectively.

1. DNA Computing Technology Iteratively Evolves

DNA computing includes DNA neural network model, DNA strand replacement network, biological probe machine and others. Basic structural unit used to construct computing model is called DNA Tile, which can perform specific function by self-programming into more complex structures according to a certain program. The basic structure stems from the studies of decidability problems. As proved, it can be used to simulate Turing Machine. Now, there are several types like cross junction, multiple strand-structure junction, single-strand tile and DNA origami tile.

Cross junction boasts the earliest-developed and most complete technology. Now, a variety of nanoscale structured grid and symmetrical 3D structure are extended, and bottom-up programming capability is gradually realized. In February 2019, a research team from California Institute of Technology (the United States) achieved binary 6-bit divisible computing by using a group of seven DNA single-tiles, controlled the molecular connectivity error rate less than 0.3‰, and proved the theoretical and practical feasibility of DNA tile computing.

Also called DNA origami, origami tile structure is the latest DNA tile structure. It can obtain random geometric patterns constructed by DNA and precisely find out their location. In recent years, remarkable progresses were made in pattern control and assembly, which reduced the complexities of pattern and coding by one order of magnitude respectively, and laid foundation for convenient coding and simple experiment. In 2019, a research team from UC Berkeley developed fully-metallized nanostructure[15] with plasma's property by using DNA origami as template, which could be widely used in the fields of imaging, labeling and sensing.

2. DNA Storing Probably Breaks the Bottleneck of Traditional Storage Performance

The coding, storage and computing of DNA molecular information are based on constructing controllable new-type nanoscale aggregate or supramolecular structure

[15] Xie et al. (2019).

by using DNA molecular property and coding theory. Its storage density is much higher than present disk media, and each piece of DNA contains 125 exabytes (1 exabyte $= 2^{60}$) of information. In 2016, University of Washington and Microsoft collaborated on developing fully-automatic system for compiling, storing an reading DNA-encoded data, which encoded a total of 739 KB data in the form of DNA sequence and stored them. Then, the system read data randomly and quickly by carefully designing specific primers and marking the address of each file on DNA sequence. In 2017, a research team from Harvard University adopted CRISPR DNA editing technology to record the image of human palm in the genome of Escherichia coli, whose reading precision reached more than 90%. In January 2019, a research team from Columbia University designed 3D-printed "DNA of Things" (DoT) storage system & structure, which integrated DNA that stored all the information of an object with the object's material to enable the object to be re-constructed from the information in any small pieces after it was damaged.

In addition to high storage-density, DNA storage possesses natural advantages in data confidentiality, integrity and non-repudiation. In November 2019, a research team from Shanghai Jiao Tong University (China) developed DNA origami cryptography (DOC) based on DNA origami technology. Whoever encrypts could convert information into lattice array and then encrypt it into framework strand that was hybridized with massive biotinylated short strands. In decryption, corresponding structure-information could be obtained by sharing the secret key of DNA framework strand and folding DNA short strands in right forms. This method integrated cryptography and steganography, and achieved *c.* 700-bit theoretical key-length by using a 7,000-basic group-long (or so) framework strand, which enormously surpassed the intensity of traditional encryption algorithm.

3. Prototypical Verification is Completed in the Cutting-edge Application of Biological Computing

Programmable self-assembled intelligent nanosystem is designed based on DNA computing principle, with great biocompatibility. It works in cells and displays advantages of high speed, sensitivity and specificity, whose nanostructure can realize the control precision at atomic level. In 2019, a research team from California Institute of Technology (the United States) developed programmable DNA computer, whose self-assembles reached 1 million DNA SSTs (Single Strand Tile). Containing computing unit sets of 355 SST-DNA molecules, it designed 21 types of programmable computing circuit modules on DNA origami base and achieved integer division computing.

### *3.3.4 Brain-Computer Interface (BCI) Progressively Develops from an Idea into Prototype Design*

Brain-Computer Interface (BCI) refers to the technology that establishes direct interactions between the brain and the external environment, which mainly includes Intrusive BCI and Non-Intrusive BCI. In the development of BCI system, hardware device, decoding algorithm and experimental paradigm all play an important part.

1. New Breakthroughs are Made in Hardware Technology

In recent years, BCI hardware develops fast. New-type nanomaterials like carbon nanotube are applied into electrode design, which triggers the devices such as electrode with anti-inflammatory coating and flexible graphene solution-gated field effect transistor for recording cortical EEG. The advance of BCI hardware technology significantly improves signal quality, extends the service life of hardware and effectively enlarges the user scale, with continuous reform and innovation in BCI field.

(1) The integration degree of BCI hardware rises gradually. In July 2019, Neuralink (the United States) released extensible high-bandwidth BCI system, which consisted of small and flexible array of electrode thread. Each array contained 96 threads and each thread contained 32 electrodes. With a total of 3,072 electrodes distributed, the system could simultaneously monitor the activities of more than 1,000 neurons and present high packing-density and high extensibility. Research teams from Kumamoto University and Yamaguchi University (Japan) integrated multi-channel measurement function of near-infrared spectrum, cortical EEG and negative temperature coefficient thermistor sensor into single device.

(2) The convenience of non-intrusive device improves remarkably. In November 2019, a team from University of Nottingham (the United Kingdom) developed wearable brain-magnetism system transformed from bicycle helmet. The system adapted to subjects of various ages, displayed high data fidelity, and permitted subject's activities. In 2020, the research team from Georgia Institute of Technology developed portable scalp electronic system based on flexible membrane circuit that used convolutional neural networks for time domain analysis and categorized steady-state visual evoked potential accurately and in real time. The research team from Tsinghua University developed EEG electrode based on multiple new materials that combined mechanical stability and flexibility, which could achieve the accuracy the same as hairless skin.

2. Achievements in Various Technological Systems are Continuously Made

In common BCI, there are two technological routes: endogenous induction paradigm and external stimulus paradigm. Endogenous induction paradigm relies on the autonomic regulation of brain rhythm, in particular, the regulation related to motion tasks. In non-intrusive method, BCI is featured with sensorimotor rhythm; in intrusive method, motion information is generally decoded from the spike potential or local field potential of neuron. External stimulus paradigm relies on external stimuli to induce the brain response, which is usually performed by event-related potential that results from visual, auditory and somatosensory stimuli.

In recent years, new BCI technological innovation continues to emerge, e.g. coadaptation BCI, emotional BCI, cognitive BCI and multi-BCI. The rise of many new technologies enlarges the scope of BCI application and further promotes the development of BCI technology. Besides, hybrid BCI that combines various advantages develops well. In 2019, the research team from University of Pittsburgh proposed hybrid BCI that comprehensively utilized information like EEG, EMG and EOG as well as other neural-activity testing methods like near-infrared spectrum and functional transcranial Doppler ultrasound.

In BCI algorithm, neural decoding algorithm receives great attention owing to its high efficiency and decoding precision. Brain signal mainly features with frequency band energy and time domain, and includes Metaheuristic method like ant colony algorithm and simulated annealing algorithm. Additionally, spatial filter can be obtained by using supervised learning and data-driven method, including Common Spatial Pattern (CSP), xDAWN and Fisher spatial filter as well as task-related component analysis. Currently, there are four types of major classification algorithms: self-adaptive classifier, matrix & tensor classifier, transfer learning and deep learning. In transfer learning, cross-individual transfer, cross-device transfer and Manifold Embedded Knowledge Transfer (MEKT) are successively proposed. In deep learning, convolutional neural network and Restricted Boltzmann Machine are used more, yet they need to improve in the support of large-scale training data sets.

3. Diversified Application of BCI is Explored

BCI application centers on decoding brain signal and outputting user's intention, and covers medical application and non-medical application. In medical application, BCI has been used in the rehabilitation training of patients with nervous system damage (e.g. cerebral stroke, amyotrophic lateral sclerosis and spinal cord injury). Its identification accuracy of perceptual signal (like language) continuously improves. In non-medical application, market-oriented development is gradually explored by combining virtual reality, smart home and other technologies.

(1) In medical application. In April 2019, the research team of UCSF used neural signals in the brain that were collected based on intracalvarium microelectrode and controlled the movements of the lip, tongue and throat, to synthesize the sound that human could effectively recognize.[16] The team also released simulated natural question-and-answer dialogue system based on high-density ECoG signal, with the decoding accuracy of generated sound and sensed sound reaching 61% and 76% respectively. It also used ECoG signal collected in text-reading and successfully trained a deep loop neural network model that could "translate" ECoG signal into continuous text, with an error rate of 3% (the

[16] Pandarinath and Ali (2019).

lowest) after test. These results gave hope to paralytic patients whose speech might be recovered.

In July 2019, the research teams from Carnegie Mellon University and University of Minnesota developed a non-intrusive mind-controlled mechanical arm that could constantly track random targets. In December 2019, the research team from Université Grenoble Alpes (France) developed a BCI system[17] that used epidural ECoG signal to control the movement of the exoskeleton system, to help quadriplegics walk again. In March 2020, Battelle Memorial Institute (the United States) released a BCI system that could restore motor and tactile functions simultaneously.[18] The system extracted tactile signal from the motor intention reflected by the activity of primary motor cortex and formed feedback to help patients with spinal cord injury recover touching perception.

(2) In non-medical application. In January 2018, Neurable, an American BCI startup, developed a brain-controlled game called *Awakening*. The game used headband with EEG electrode and VR headset device so that wearers could train their brains to interact with brain signals. In November 2018, the research team from UC Berkeley developed portable steady-state visual evoked potential BCI that combined VR headset device.[19] In April 2019, the research team from Chinese Academy of Sciences developed a code modulation-based visual evoked potential that could be used to achieve the recognition of individual identity by utilizing the non-mimetic nature of EEG biological information.[20]

## 3.4 New Technologies and New Applications

With the development of new technologies like AI, big data and 5G, many technological enterprises provide differentiated new products & solutions, forge diverse new application ecosystems and boost the rapid growth of smart economy, according to specific use cases. In both products and services, new technologies integrate with traditional industries quickly and fully develop in various fields like medical treatment, transportation, home furnishing and manufacturing.

### *3.4.1 WITMED Realizes Rapid Development*

WITMED refers to realizing the interaction between patients and medical staff, institutions and devices and gradually achieving informatization by establishing

[17] Benabid et al. (2019).

[18] Ganzer et al. (2020).

[19] Abbasi-Asl et al. (2019).

[20] Zhao et al. (2019).

comprehensive medical information platform and combining the most advanced IoT technology. In 2016–2018, global WITMED service expenditure grew at a CAGR (Compound Annual Growth Rate) of around 60%, with the industrial scale reaching $400 billion in 2019 as estimated.[21]

1. WITMED Model Changes Traditional Medical Treatment

(1) Health management model. As modern electronic devices like mobile phone become popular, the efficiency of health management improves remarkably and the cost decreases significantly, which fosters the new-type health management system.
(2) Models in registration, consultation and medicine-purchase. Formerly, patients saw doctors and bought medicines with doctors' lists. Now, critical patients make online appointments, and common patients see doctors online, such as online consultation and online medicine-purchase.
(3) Doctor-patient ecology. On-line diagnosis and treatment can evidently reduce direct doctor-patient contact and ease doctor-patient tension without affecting information exchange.
(4) The application of intelligent medical facilities. AI-technology-based intelligent software and device like Clinical Decision Support System (CDSS) and AI Fundus Screening System can help frontline doctors better complete their diagnosis. Besides, when major public health events happen, the deployment and use of intelligent medical facilities can effectively improve the prevention and treatment of sudden and epidemic diseases. Taking the fight against the COVID-19 Pandemic in 2020 as an example. Intelligent online diagnostic systems in various countries connected wards and healthcare centers simultaneously, and medical experts treated patients without entering quarantine areas, which prevented medical staff from being infected by the virus.

2. The Use Cases of WITMED

(1) Image-aided diagnosis. In July 2019, QuantX, an American AI assistance system, was approved by U.S. Food and Drug Administration. QuantX used AI software to assist radiologists and analyze Magnetic Resonance Imaging (MRI) in order to confirm or question doctor's diagnoses. As clinical studies proved, QuantX could reduce the missed diagnosis of cancer by 39% and increase the overall accuracy by 20%. In June 2020, China's BioMind, an intracranial tumor MRI-aid diagnostic software, was approved by National Medical Products Administration (NMPA) for Class-III medical devices.
(2) Disease-risk prediction. Disease-risk can be predicted in advance through gene sequencing and testing. Or the onset of disease can be predicted by using various big data on biochemistry, imaging and daily behavior. In May 2020, DeepMind and Google Health, two AI companies, developed new AI medical system to detect age-related macular degeneration (AMD). The system used

[21] Baidu and China Development Research Foundation, *New Infrastructure Construction, New Opportunities: White Paper on the Development of China's Smart Economy*, June 2020.

two neural networks to analyze 3D eye-scanning results and mark information training and learning of 2,795 cases. Its prediction on the condition of early-stage patient's in next 6 months accorded with the judgment of experts, but displayed much higher efficiency than manual model.

### 3.4.2 *Intelligent Traffic App Gradually Spreads*

Intelligent traffic applies advanced technologies like AI, information communication and sensing & control into ground traffic management system, to achieve large scale, all-round, real-time, accurate and efficient traffic management. As public data suggests, intelligent traffic can improve the efficiency of road use, reduce traffic congestion by around 60% and increase the traffic capacity of the existing roads by 200–300%. On condition that vehicles run in the intelligent traffic control system, their parking time can decrease by 30% and running time by 15%-45%, with vehicle operation efficiency up by over 50%.[22] By 2025, global intelligent traffic market scale will reach $262.1 billion as predicted, with a CAGR (Compound Annual Growth Rate) of 18.68%.[23] As intelligent traffic market gradually matures, the industry turns from foundation-building to high-quality development.

1. Four Major Trends in the Development of Intelligent Traffic

(1) Autonomous driving.
(2) Big data-based MaaS one-stop travel service.
(3) Control of traffic cloud brain and intelligent road network in road system.
(4) Proactive prevention and control of traffic safety, intelligent operation & maintenance and sound management of rail transit, delicacy governance of urban parking, and autonomous traffic system.

Among them, as an important part of intelligent traffic control system, autonomous driving vehicle plays a major role in achieving vehicle–road collaborations and improving citizen's travel experience. Traditional automobile manufacturers and emerging Internet enterprises gradually promote the realization of L3 level of autonomous driving with their own advantages in automobile safety technology and intelligent technology. So far, more than 25 cities globally (countries and regions) test autonomous driving cars.

2. The Use Cases of AI in Traffic Become More Extensive

(1) In autonomous driving. AI is mainly applied into vehicle's autonomous driving model, from vehicle perception to decision as well as customized prediction & maintenance function. This can increase mobility, reduce the incidence of traffic accident and diminish the demand for urban parking lots.

---

[22] Baidu and China Development Research Foundation, *New Infrastructure Construction, New Opportunities: White Paper on the Development of China's Smart Economy*, June 2020.

[23] Smart Thermostat Market-Growth, Trends, and Forecast (2020—2025), Mordor Intelligence.

(2) In urban traffic. With the support of AI software and hardware system, sensor, imaging system and traffic remote communication & monitoring system, real-time traffic condition can be obtained, based on which traffic signs can be changed, in order to reduce traffic jam and carbon emission. This improves pedestrian's safety and raises the quality of life.
(3) In parking. With the help of AI and cloud-end data analysis, apps can be activated to choose roadmap and match parking lot, in order to recognize vacant parking lots and inform vehicles of them.
(4) In expressway. AI is gradually applied and developed in IoV & unmanned driving, emergency plan matching, non-sensory payment, fee evasion inspection, intelligent interactive customer service, behavior supervision, intelligent path planning and traffic guidance.

### *3.4.3 Smart Home Develops in a Diverse Way and Competition Among Enterprises Intensifies*

Smart home uses AI and IoT technologies to integrate facility related to household life, construct efficient residential facility and family daily-affair management system, improve the safety, convenience, comfort and art at home, and build environmental-friendly and energy-saving living environment. Smart home frees citizens from traditional home appliances model and stresses intelligent and automatic use.

1. Global Smart Home Market Grows Rapidly

According to the data published by Strategy Analytics, a market research company, in 2019, consumers expended as much as $103 billion on smart home-related hardware, service and installation, which will reach $157 billion in 2023, with a CAGR (Compound Annual Growth Rate) of 11%. Multimedia entertainment facilities like smart loudspeaker and smart TV, as well as lighting-control device and home monitoring & security system, will take up the largest share of smart home market. Amazon, Apple, Google, Samsung, ADT, Honeywell, Bosch, Assa Abloy, ABB, Ingersoll Rand and General Electric dominate global smart home industry, accounting for 40–45% of global smart home market share.[24]

2. The Use Case of Smart Home

On the one hand, smart home uses sound interface as app access, and on the other, it constructs comprehensive ecological system by integrating household appliances. Smart loudspeaker is the practical product of sound interface, as a tool by which family consumers use sound to access Internet. For example, it can be used to play songs on demand, shop online, or watch weather forecast. It can also be used to control

[24] Fortune BusinessInsights, Smart Home Market Size, Share & Industry Analysis 2019–2026[ol]. https://www.fortunebusinessinsights.com/industry-reports/smart-home-market- 101,900.

smart home devices, e.g. opening curtain, setting the temperature of refrigerator and heating up water heater in advance. In 2019, sound assistants, such as Google Assistant, Apple's Siri and Amazon's Alexa, became a technological hot topic. As sound assistant is implanted into more and more external software and hardware products, consumers are allowed to easily control home devices with their sounds. Besides, Google Home and Amazon's Alexa allow the third party to enter their ecosystem of sound services.

### 3.4.4 The Industrial Upgrading of Intelligent Manufacturing Accelerates

Intelligent manufacturing deeply integrates new-generation information communication, AI technology and advanced manufacturing technology, and helps manufacturing industry to upgrade from mechanization, electrification and automation to digitization, interconnectivity and intelligentialization. Specifically, digitization means to transform industrial information into digital format by computer management. Interconnectivity corresponds to Internet of Everything (IoE) and achieves producer-machine, machine-machine and consumer-producer connection. Intelligentialization means to realize the free flow of data and the intelligent decision-making for various cases via big data analysis and AI technology.

Intelligent manufacturing becomes an important developmental trend in manufacturing industry. As related statistics[25] indicates, the scale of global intelligent manufacturing market would reach $214.7 billion in 2020 and $384.8 billion in 2025, with a CAGR (Compound Annual Growth Rate) of 12.4%. As 3D printing, simulation analysis, industrial IoT and other technologies penetrate into manufacturing industry, automobile, aerospace and national defense firstly achieve growth in the field of intelligent manufacturing, and energy & equipment manufacturing industry keeps high-speed growth. Globally, countries can be categorized into four echelons in intelligent manufacturing: The first echelon is leading countries that control advanced technology, patent and brand, represented by the United States, Japan and Germany. The second echelon is advanced countries that can produce key components with some core technologies and large-scale integration capabilities, represented by China, South Korea, the United Kingdom and Sweden. The third echelon is potential-type countries that own less core technologies and concentrate on parts processing. The fourth echelon is lagging countries that provide raw materials and develop labor-intensive manufacturing.

[25] Data Source: Development Research Center of the State Council, *New Infrastructure Construction, New Opportunities: Development of China's Smart Economy*, June 2020.

## References

Abbasi-Asl, R., Keshavarzi, M., & Chan, D. Y. (2019). Brain-computer interface in virtual reality, in *2019 9th International IEEE/EMBS Conference on Neural Engineering (NER).* (pp. 1220–1224). IEEE.

Benabid, A. L., Costecalde, T., Eliseyev, A., et al. (2019). An exoskeleton controlled by an epidural wireless brain–machine interface in a tetraplegic patient: A proof-of-concept demonstration. *The Lancet Neurology, 18*(12), 1112–1122.

Chen, Z., Tagliasacchi, A., & Zhang, H. (2020). Bsp-net: Generating compact meshes via binary space partitioning, in *Proceedings of the IEEE/CVF Conference on Computer Vision and Pattern Recognition* (pp. 45–54).

Ganzer, P. D., Colachis 4th, S. C., Schwemmer, M. A., et al. (2020). Restoring the sense of touch using a sensorimotor demultiplexing neural interface. *Cell.*

Habermann, M., Xu, W., Zollhofer, M., et al. (2020). Deepcap: Monocular human performance capture using weak supervision, in *Proceedings of the IEEE/CVF Conference on Computer Vision and Pattern Recognition* (pp. 5052–5063).

Lan, Z., Chen, M., Goodman, S., et al. (2019). ALBERT: A Lite BERT for self-supervised learning of language representations, in *International Conference on Learning Representations.*

Pandarinath, C., & Ali, Y. H. (2019). Brain implants that let you speak your mind. *Nature, 568*(7753), 466–468.

Wu, Z., Chen, Y., Kao, B., et al. (2020). Perturbed masking: parameter-free probing for analyzing and interpreting BERT. arXiv:2004.14786.

Wu, S., Rupprecht, C., Vedaldi, A. (2020). Unsupervised learning of probably symmetric deformable 3D objects from images in the wild, in *Proceedings of the IEEE/CVF Conference on Computer Vision and Pattern Recognition* (pp. 1–10).

Xie, N., Liu, S., Fang, H., et al. (2019). Three-dimensional molecular transfer from DNA nanocages to inner gold nanoparticle surfaces. *ACS Nano, 13*(4), 4174–4182.

Zhao, H., Wang, Y., Liu, Z., et al. (2019). Individual identification based on code-modulated visual-evoked potentials. *IEEE Transactions on Information Forensics and Security, 14*(12), 3206–3216.

# Chapter 4
# The Development of World Digital Economy

## 4.1 Outline

Impacted by the tension in global trade and other factors, the world economy grew by 2.9% in 2019, reaching the lowest level in last decade. Since early 2020, the COVID-19 Pandemic spread globally and lashed global production, trade, cross-border investment and financial market, which intensified the risk of global economic downturn. International Monetary Fund (IMF) predicted that the world economy would grow by -4.9% in 2020. Digital economy becomes an important force in hedging against the impact of the COVID-19 Pandemic, re-shaping economic pattern and improving governance capacity. Major countries in the world continuously strengthen the strategic layout of digital economy and gradually improve the strategic design and policy system. The United States and China rank the first and the second in the world in digital economy scale. European countries develop slowly in digital economy, and emerging countries display great potential in the development of digital economy.

Global digital industrialization develops steadily, basic telecommunications market slows down, and electronic information manufacturing industry owns great potential. IT service industry like big data, AI and blockchain keeps thriving, and Internet information content service industry develops quickly. Industrial digitization deeply advances, and the digitization of manufacturing industry further deepens, with remarkable features of cross-border integration, platformization and sharing. The digitization of service industry quickens in transformation and upgrading, and the digitization of agriculture develops stably. Fintech flourishes, with the emphasis on virtual currency, digital bank and digital-financial regulation. E-commerce continues to expand, emerging market grows fast, and mobile e-commerce hits new highs.

In a short time, some countries that highly rely on external economy start to re-consider their industrial policies in order to reduce their over-dependence on external world. In a long time, new business forms, new models and new industries that develop rapidly in the COVID-19 Pandemic will forge new growth point for economic globalization. Featured with digitization, network and intelligentization,



Chinese Academy of Cyberspace Studies, *World Internet Development Report 2020*,
https://doi.org/10.1007/978-981-16-9388-5_4

new technologies develop in an innovative way and their applications accelerate. Both developed countries and emerging economies center on global digital transformation to promote domestic industrial construction and regional development. Global supply chain, industrial chain, service chain and value chain will form closer connection in reconstruction and advance economic globalization in newer form and at higher level.

## 4.2 The Developmental Trend of World Digital Economy

The new-round scientific & technological revolution and industrial transformation re-shape global economic pattern fast. Innovation and application of information technology emerge continually. As a new-type productive factor, data takes global digital economy to a new level. Digital trade thrives, when the COVID-19 Pandemic speeds up the digital transformation of global economy.

### *4.2.1 The Developmental Strategy Becomes More Targeted*

Major countries in the world take the development of digital economy as an important measure to increase the quality and efficiency of real economy and re-forge their core competitiveness. They successively formulate developmental strategies on digital economy and boost the digital transformation of economy and society.

1. Technologies like 5G, Blockchain and AI remain strategic priorities

In 5G, in January 2020, U.S. House of Representatives voted to pass the *Promoting United States International Leadership in 5G Act* and *Promoting United States Wireless Leadership Act*, which clarified that the United States and its allies & partners should take part in and lead international organizations that established standards on 5G and next-generation mobile telecommunication system and infrastructure. In April 2020, Russian Ministry of Digital Development, Communications and Mass Media drafted new strategic document on the development of 5G network, which ruled that four telecom operators would divide regions respectively to achieve the exclusive deployment of 5G network. In June 2020, South Korea announced *Economic Stimulated Plan* with $63 billion, which strived to foster national economic growth by increasing investment and conducting the systemic reform, to promote the use of 5G and AI among various industries and to advance digitization in its most undeveloped regions.

In blockchain, in September 2019, German Federal Government released *German National Blockchain Strategy*, which aimed to construct systematic framework for the innovation and development of blockchain in Germany, to guide the reasonable development of blockchain technology, to promote the prosperity of German token economy, and to tap the potential of blockchain in strengthening the digital transformation of economy and society. In February 2020, Australian government issued *National Blockchain Roadmap*, which included the "regulation, skill and capacity building, innovation, investment, and international competitiveness and cooperation" of blockchain and aimed to promote Australia to be a global leader in blockchain industry.

In AI, in November 2019, U.S. Congressional Research Service (CRS) updated its research report *Artificial Intelligence and National Security*. The updated edition mainly centered on eight aspects, involving AI terminology and background, the problems that U.S. Congress faced, AI application in national defense, overall challenges for military AI, international competitors, international organizations, opportunities and challenges from AI and AI's impact on war. In November 2019, Singapore launched *National Artificial Intelligence Strategy* to widely popularize the use of AI technology in five major fields, i.e. transport and logistics, smart town and neighborhood, healthcare, education and security & safety. Singapore government invested 500 million Singapore dollars to deepen AI research, innovation and enterprise activities, under "Research, Innovation and Enterprise 2020" Plan. In March 2020, European Commission published *White Paper on Artificial Intelligence* and *European Data Strategy*, which respectively proposed to establish "Trustable Artificial Intelligence Framework" and truly unified data market. As planned, EU's share in data economy would basically stand comparison with its economic weight by 2030, with the vision that EU would become the most attractive, secure and dynamic data-agile economy in the world.

2. To promote the comprehensive digital transformation of economy and society constitutes strategic direction

In March 2020, European Commission announced a series of reports, such as *Shaping Europe's Digital Future*, *A European Industrial Strategy*, *Small and Medium-Sized Enterprises Strategy for Sustainable and Digital Europe*, *Single Market in Service for European Businesses and Consumers* and *Identifying and Solving Barriers in Single Market*. It also presented a comprehensive action plan for the future, which covered formulating intellectual property right action plan, protecting technological ownership and strengthening of European industrial and strategic autonomy. These aimed to help the European industry complete digital transformation and improve competitiveness and strategic autonomy. In May 2020, president of European Commission submitted "Next Generation EU" Recovery Plan to European Parliament, which proposed to establish a special recovery fund totaling 750 billion euros and forge green new deal, digital transformation and resilience to crisis. In June 2020, Vietnam released *National Digital Transformation Plan by 2025 & Developmental Direction by 2030*, which proposed that by 2025, digital economy would account for 20% of Vietnam's GDP and at least 10% of all industries and fields, that information development index and global cybersecurity index would rank top 50 in the world, and that global innovation index would rank top 35; by 2030, Vietnam would popularize fiber-optic broadband and 5G mobile network service, with over 80% population opening electronic payment account.

### 4.2.2 The Developmental Pattern Basically Maintains Stable

The explosive growth of digital economy accelerates the reconstruction of global economic pattern. China and the United States take the lead in the development of world digital economy, and newly developing countries come to rise.

1. China and the United States Take the Lead in the Development of World Digital Economy According to the report *New Prospects of Global Digital Economy (2019): Accelerating Take-off and Restoring Growth* published by China Academy of Information and Communications Technology, in 2018, the United States remained the first in the world in digital economy scale, reaching $12.34 trillion; China realized around $4.73 trillion in digital economy scale and maintained the position of the second largest economy in the world. Despite the gap in Chinese and American digital economy scale, the gap in the competitiveness of two digital economies gradually narrows. According to *Global Digital Economy Competitiveness & Development Report (2019)* compiled by Shanghai Academy of Social Sciences, in national competitiveness, the United States (75.94), Singapore (60.96) and China (57.37) rank the first, the second and the third in World Digital Economy National Competitiveness List, followed by the United Kingdom(51.61), Finland (50.11), South Korea (49.86) and Japan (49.51). The gap in the competitiveness of digital economy between China and the United States gradually narrows, from 23.82 points in 2016 to 18.57 points in 2018. China surpasses the United States and ranks the first in the world in the competitiveness of digital industry, yet lacks competitiveness in the governance of digital economy.

2. Digital economy develops slowly in Europe

According to *Digital Economy and Society Index (DESI) 2020* published by European Commission, in overall digital level, Finland, Sweden, Denmark and the Netherlands achieved the highest digital level, and Bulgaria and Greece stayed at the bottom. In the past five years, Ireland rose the most quickly, followed by the Netherlands, Malta and Spain. In the last decade, the development of information technology gave great potential to digital economy. However, European Union performed not well in the field of digital economy. As the data of the World Bank suggested, in 2019, EU's economic aggregate accounted for *c.* 15.77% of the world; yet, European digital enterprises accounted for less than 4% of the market value of global digital enterprises. As the data of CB Insights showed, by June 2020, the number of unicorns in the United States totaled 228, ranking the first in the world; the number in China totaled 122, ranking the second in the world; the number in the United Kingdom totaled 25 and surpassed Germany, ranking the third in the world. India, Germany and South Korea owned 21, 13 and 10 unicorns respectively.

3. Emerging countries display enormous potential in digital economy

In 2019, supported by the World Bank, African Union launched "Digital Moon" Initiative, which aimed to give high-speed connectivity to all citizens in African continent and establish solid foundation for active digital economy, with 1.65 billion to 1.71 billion Africans digitally connected by 2030. Mobile payment becomes a hot topic in the competition of emerging market. Take India as an example. According to the report released by World Line India, mobile payment market in India hits a new high, with Bangalore as the leading city in digital transaction, followed by Chennai, Mumbai and Poona. Facebook, Amazon, Walmart, Alphabet, Microsoft, Berkshire Hathaway and Alibaba have all entered Indian market and competed fiercely.

### 4.2.3 Internet Enterprises Rise Rapidly

In "Top 100 Corporations in Global Digital Economy List 2019" released by Forbes, the United States had 38 listed corporations, with the largest number. China ranked the second, with 14 listed corporations. In the "Top 10 Corporations List", there were eight American corporations, with Apple ranking the first. According to the data on corporate financial report, in 2019, Amazon, Apple and Samsung ranked top three in business revenue in global IT corporations, and China's Huawei and JD ranked the 7th and the 9th respectively. In net profit, in the "Top 10 Global IT Corporations in Net Profit 2019", there were 7 American corporations, and South Korea's Samsung, China's Tencent and Alibaba ranked the 6th, 7th and 8th respectively. See Table 4.1 for "List of Top 10 Global IT Corporations in Business Revenue and Profit in 2019 (FY)".

**Table 4.1** List of top 10 global IT corporations in business revenue and profit in 2019 (FY)

| Ranking | Corporation | Business revenue | Year-on-year growth (%) | Ranking | Corporation | Net profit | Year-on-year growth (%) |
|---|---|---|---|---|---|---|---|
| 1 | Amazon | 2805.22 | 20.45 | 1 | Apple | 552.56 | –7.18 |
| 2 | Apple | 2601.74 | –2.04 | 2 | Microsoft | 392.40 | 136.80 |
| 3 | Samsung | 1956.00 | –11.72 | 3 | Google | 343.43 | 11.74 |
| 4 | Hon Hai | 1776.67 | 3.06 | 4 | Intel | 210.48 | –0.02 |
| 5 | Google | 1618.57 | 18.30 | 5 | Facebook | 184.85 | –16.40 |
| 6 | Microsoft | 1258.43 | 14.03 | 6 | Samsung | 184.56 | –54.21 |
| 7 | Huawei | 1217.77 | 15.77 | 7 | Tencent | 133.75 | 25.30 |
| 8 | Dell | 906.21 | 14.65 | 8 | Alibaba | 119.55 | 17.19 |
| 9 | JD | 828.64 | 23.31 | 9 | Cisco | 116.21 | 10,464.55 |
| 10 | Sony | 781.40 | 1.42 | 10 | Amazon | 115.88 | 15.04 |

*Data Source* 1. http://news.mydrivers.com; 2. Data on Corporate Financial Reports
*Unit* USD 100 Million

### 4.2.4 *Investment Markets in Chinese and American Digital Economy Thrive*

The United States and China see the most active markets in the Internet investment and financing in the world, remaining in the first tier for a long time. As the data of CB Insights suggested, in the second quarter of 2020, global financing deal totaled 3,812, down by 9% year on year (2019). Regionally, financing deal in North America was the most, reaching 1,479 and accounting for 38.8%. Asia followed North America, accounting for 35.4%. Europe accounted for 22.1% and other regions accounted for 3.7%. In the financing amount, global financing totaled $50.2 billion, down by 13% year on year (2019). In specific, North America, Asia and Europe accounted for 97.6% together, reaching $49 billion. In top 6 corporations in financing amount, four came from the United States, and two came from China and the United Kingdom. In three cases, financing amount exceeded $500 million, i.e. Zuo Ye Bang, a Chinese online education platform ($750 million); Stripe, a U.S. digital payment platform ($600 million); and Palantir Technologies, a U.S. big data corporation ($500 million). In financing sector, emerging sectors take up larger proportion in venture capital. Digital healthcare, Fintech and AI dominate and medical device and cybersecurity follow them. Specifically, 173 financing cases happened in digital healthcare, totaling $3.1 billion in amount. This remained invariant compared with the first quarter of 2020, with an increase in number.

### 4.2.5 *Global Digital Trade: Development & Potential Danger*

ICT promotes the upgrading and transformation of traditional goods-trade model. Digital trade plays an increasingly important role in global trade. Thriving digital trade triggers discussions and gaming on the digital tax in the world.

1. Digital service trade[1] grows rapidly

As *White Paper on the Development and Influence of Digital Trade (2019)* released by China Academy of Information and Communications Technology suggested, in 2008–2018, the export scale of global digital delivery trade increased from $1,837.99 billion to $2,931.4 billion, up by around 60%, with a Compound Annual Growth Rate (CAGR) of around 4.78% (3.80% in the export of service trade, 1.87% in the export of goods trade in the corresponding period). Its share in the export of service trade increased from 45.66 to 50.15%.

2. The advantages of developed countries further enlarge

Compared with other trade models, the gap between developed countries and developing countries in digital service trade further enlarges. According to *White Paper*

---

[1] Digital service refers to product and service that can be delivered remotely via Internet.

*on the Development and Influence of Digital Trade (2019)*, in 2018, international market share of developed economies in digital service trade, service trade and goods trade reached 76.1%, 67.9% and 52% respectively. In specific, the export scale of digital service in the United States and European Union reached $466.72 billion and $1,449.06 billion respectively in 2018, accounting for more than 65% of world digital service export. In China and India, two largest developing countries, the export scale of digital service reached $131.45 billion and $132.6 billion, accounting for only 4.45% and 4.52% respectively in the world.

3. Digital tax arouses global concern

The existing tax regulation and collection framework do not accord with the robust development of digital economy. This arouses high concern in various countries on setting up rules for global digital trade. Since 2018, some countries have proposed GAFA tax (named after the first English letters of Google, Apple, Facebook and Amazon), mainly levying a 3% sale-tax on the sales amounts of transnational giants in their home countries. British government announced that from April 1, 2020, digital service tax would be levied. Enterprises that provided British users with services on social media, search engine and online marketing must be levied by 2% of the surplus as digital service tax on condition that they obtained global revenue of more than £500 million from digital services and that the revenue from British users exceeded £25 million. Organization for Economic Co-operation and Development (OECD) initiated and led the research on international tax reform in economic digitization. OECD's reform proposal gave countries where the markets and users were formed legal rights to partake in the allocation of transnational corporation's income tax. The reform on digital tax requires to adjust international rules closely related to the interests of all countries. Under the background of the rapid spread of the COVID-19 Pandemic in the world, major countries continually debate over digital tax and undergo multi-party negotiation and arduous gaming.

4. Challenge and opportunity co-exist in the development of digital economy

Under the impact of the COVID-19 Pandemic, both supply and demand in major global markets face pressures, and both internal and external influences overlap. International supply chain and supply–demand end of core component and high-end product shrink, which lashes the industrial chain of world digital economy. In some countries and regions, upstream and downstream industrial chains suffer the risk of rupture, which poses great pressure on enterprise's follow-up production & business and marketing. Besides, some countries uphold unilateralism and intensify global market turbulence. In the long term, the development and innovation & application of new technologies like big data, AI and blockchain will impel developed countries and emerging economies to center on digital transformation and advance national industrial development. Closer connection will be established among global supply chain, industrial chain, service chain and value chain in reconstruction, and economic globalization will be promoted to a higher level in new form.

### 4.2.6 *Digital Economy Contributes to Global Fight Against the COVID-19 Pandemic*

The COVID-19 Pandemic severely impacted on global economy. As Asian Development Bank estimated, the COVID-19 Pandemic would result in economic loss of as much as $8.8 trillion globally. Kristalina Georgieva, chair of International Monetary Fund (IMF), stated that global economy would suffer a loss of $12 trillion due to the impact of the COVID-19 Pandemic. In the research report released by United Nations Conference on Trade and Development (UNCTAD), it was mentioned that global crisis caused by the COVID-19 Pandemic further fostered the development of digital economy and exerted lasting influence after the recovery of global economy. As the COVID-19 Pandemic spread globally, various countries in the world successively used digital technology to hedge against the impact of the COVID-19 Pandemic and spurred the digital transformation of global economy. Schools in Italy, South Korea, the United Kingdom and other countries suspended and students switched to online learning. Corporations like Apple, Google and Facebook required employees to work at home. In the second quarter of 2020, Zoom Video Communications, a developer of online video conferencing software, realized a total revenue of $663.5 million, up by 355% year on year. In the COVID-19 Pandemic, numerous enterprises of digital economy took advantages of information technology and data and played a unique role in supporting the prevention & control of the COVID-19 Pandemic, stimulating consumption and stabilizing growth, helping the resumption of work and advancing production, benefitting people's livelihood and ensuring stability, and improving governance capacity.

## 4.3 Digital Industrialization Grows Steadily

In 2020, the development of global basic telecom industry developed stably, with a slowdown in market growth. Generally speaking, electronic information manufacturing stagnated, and wearable device and 5G smartphone presented great potential in growth. New-generation IT service industry like big data, AI and blockchain kept thriving. Internet information content service industry developed quickly and became a new force in digital industrialization.

### 4.3.1 *Basic Telecommunications Progress Stably*

Global basic telecommunications industry grows steadily. The COVID-19 Pandemic stimulates more demand for sound service, 5G drives the development of telecom business, and satellite communications industry further advances toward a stage of deep adjustment.

1. Telecom service industry shows a favorable trend

As the report of Grand View Research suggested, in 2019, global telecom service realized a business revenue of around $1.74 trillion, growing with an estimated Compound Annual Growth Rate (CAGR) of 5.0% in 2020–2027. Impacted by the COVID-19 Pandemic, the demands for apps like digital entertainment, teleworking and social media continuously grew, which resulted in the jumping of business in sound service. With the help of large netizen number and high smartphone penetration rate, China, Japan and India are major contributors to the growth of regional market. As Global System for Mobile Communications Association (GSMA) predicts, the Asia–Pacific Region probably attracts more than half of all new mobile users by 2025.

2. Global 5G market probably achieves high-speed growth

According to the report of Omdia, an international consultancy, by the end of 2019, the number of global 5G commercial network reached 62, and cross-regional 5G network coverage grew remarkably. In such countries as China, South Korea, Australia, Germany, the United Kingdom and the United States, 5G market continues to develop. As the data of Grand View Research, a market research agency, indicated, in 2019, global 5G infrastructure market realized a business revenue of only $1.9 billion and would grow rapidly in 2020, with a Compound Annual Growth Rate (CAGR) of 106.4% in 2020–2027. According to the data of Global System for Mobile Communications Association (GSMA), 20.0% of global communications connectivity will be united by 5G network by 2025.

3. Revenue in satellite communications business grows

In 2019, satellite communications industry further formed new pattern of deep adjustment of revenue structure and continuous change in business application. The revenue of traditional businesses represented by TV broadcast and fixed communication continued to slow down or even declined, yet these of data & network businesses continued to grow, with fierce competition in global satellite operation service. According to the data released by Grand View Research, a market research agency, in 2019, global satellite data service market realized a scale of $5.3 billion, with a Compound Annual Growth Rate (CAGR) of over 27% in 2020–2027, as predicted.

### *4.3.2 Electronics and Information Manufacturing Industry Displays Great Potential*

In 2019, electronic information industry stagnated yet presented great growth potential, particularly in semiconductor industry. Wearable device and smartphone industries that were closely associated with mass consumers flourished.

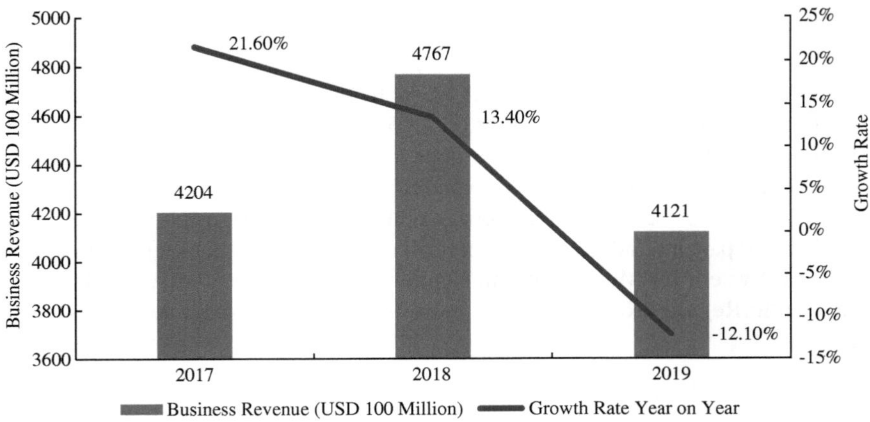


**Fig. 4.1** The Trend of Global Semiconductor Business Revenue in 2017–2019

1. Global semiconductor market relatively stagnates

According to *2019 Global Semiconductor Market Report* released by Semiconductor Industry Association (SIA), global semiconductor industry displayed a downturn trend, with annual business revenue of $412.1 billion, down by 12.1% year on year (2018), due to the lower prices of flash memory and memory. In December 2019, global semiconductor sales reached $36.1 billion, down by 5.5% year on year. In specific, the revenue in American market was $7.49 billion, down by 10.8% year on year; the revenue in European market was $3.2 billion, down by 7.6% year on year; the revenue in Japanese market was $3.04 billion, down by 8.3% year over year; the revenue in Chinese market was $12.81 billion, up by 0.8% year on year; the revenue in the Asia–Pacific Region (excluding China and Japan) was $9.56 billion, down by 7.5% year on year. See Fig. 4.1 for "The Trend of Global Semiconductor Business Revenue in 2017–2019".

2. 5G becomes new growth point for smartphone

According to the data released by International Data Corporation, in 2019, global smartphone shipment reached 1.371 billion, down by 2.3% year on year for three successive years. In mobile phone manufacturers, Samsung, Huawei, Apple, Xiaomi and OPPO ranked top five in shipment in 2019, 296 million, 241 million, 191 million, 126 million and 114 million respectively. In the market share of shipment, top five mobile phone manufacturers accounted for more than 70%. Among them, Samsung took up 21.6%, Huawei took up 17.6%, Apple took up 13.9%, Xiaomi took up 9.2% and OPPO took up 8.3%, in market share.

As International Data Corporation predicted, in 2020, global shipment of 5G mobile phones would reach 190 million, which accounted for 14% of the total smartphone shipment. According to the data and research on supply chain released by China Mobile, the scale of 5G mobile phones in China would reach 150 million in 2020.

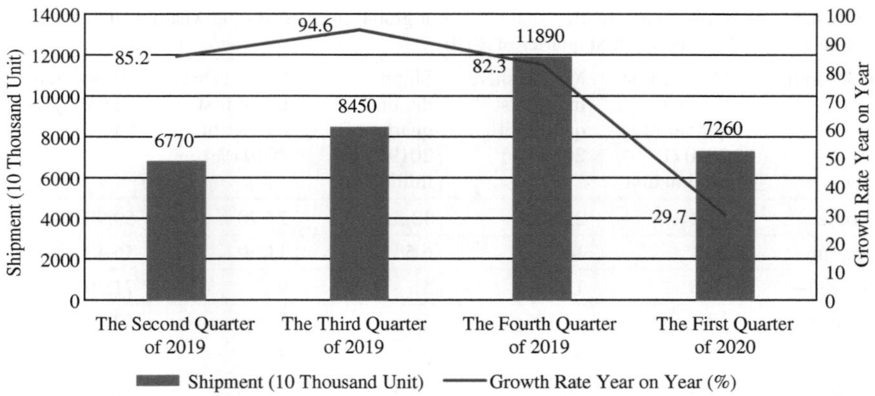


**Fig. 4.2** Comparison of the shipments of global wearable devices in the last three quarters of 2019 and the first quarter of 2020

3. Wearable device market grows fast

According to data published by International Data Corporation, in the first quarter of 2020, global shipment of wearable device reached 72.6 million, up by 29.7% year on year (see Fig. 4.2). The shipment of wearable device bucked the trend and grew, which benefitted from the rapid growth of wristband and earphone devices. In specific, in the first quarter of 2020, the wristband grew by 16.2% and earphone device grew by 68.3%, accounting for 54.9% of the total market.

In the market share of major corporations, in the first quarter of 2020, Apple completed 21.2 million in shipment, accounting for 29.3% of the market. Xiaomi followed closely, with a shipment of 10.1 million and a proportion of 14%. Samsung ranked the third, with a shipment of 8.6 million; in specific, earphone business took up 74% of total shipment, up by 15.1% year on year. Huawei and Fitbit ranked the fourth and the fifth respectively. See Table 4.2 for "The Shipment and Market Share of Major Global Corporations on Wearable Devices in the First Quarter of 2019 and the First Quarter of 2020".

### *4.3.3 Software and Information Technology Service Industry Advances Smoothly*

In 2020, generally speaking, software and information technology service industry developed steadily. Global data volume grew rapidly, which promoted the sustainable and vigorous development of big data industry. After three-year fast development, the growth rate of global blockchain industry presented a downturn trend yet robust potential. AI drove the continuous development of software and information technology service industry. Global technological giants successively deployed the industry to control user access and industrial leading position.

**Table 4.2** The shipment and market share of major global corporations on wearable devices in the first quarter of 2019 and the first quarter of 2020

| Corporation | Shipment in the first quarter of 2020 /1 million unit | Market share in the first quarter of 2020 (%) | Shipment in the first quarter of 2019 /1 million unit | Market share in the first quarter of 2019 (%) | Growth rate year on year (%) |
|---|---|---|---|---|---|
| Apple | 21.2 | 29.30 | 13.3 | 23.70 | 59.90 |
| Xiaomi | 10.1 | 14 | 6.5 | 11.60 | 56.40 |
| Samsung | 8.6 | 11.90 | 5 | 9 | 71.70 |
| Huawei | 8.1 | 11.10 | 5 | 8.90 | 62.20 |
| Fitbit | 2.2 | 3 | 2.9 | 5.20 | –26.10 |
| Others | 22.3 | 30.80 | 23.3 | 41.60 | –4 |
| Total | 72.6 | 100 | 56 | 100 | 29.70 |

1. Big data industry keeps thriving

Globally, intelligent terminal becomes more and more popular, which promotes the fast development of big data industry. As Market Watch, an international consultancy, estimated, global big data market realized a business revenue of around $54.22 billion in 2020, which probably would reach $156.72 billion by 2026, with an average annual growth rate of 19.3%.

Regionally, in big data service market scale, North American boasts great advantages, with leading corporations like IBM, Oracle, Microsoft, HP, SAP and Amazon. According to the statistics of IBM and Prospective Industrial Research Institute, the United States owns the largest number of data centers, accounting for 44%. China, Japan, the United Kingdom, Australia and Germany account for 8%, 6%, 6%, 5% and 5% respectively. In construction & development, the United States remains the role of market leader and guides the development of global data center market in data center product, technology and standard. With the help of netizen number, the market in the Asia–Pacific Region grows the fastest. Compared with the corresponding period in 2018, the growth rate of investment reached 12.3%, and the scale of investment reached $75.17 billion. The steadfast development of Chinese data center market further promoted the sustainable and rapid growth of big data market scale.

2. The competition in cloud computing market intensifies

In 2019, the competition in global cloud computing market became fiercer. As the data released by Canalys suggested, in 2019, global cloud computing market scale firstly exceeded $100 billion and reached $107.1 billion, jumping by 37.6% year on year. Amazon realized a business revenue of $34.6 billion in cloud computing and ranked the first, up by 36.0% year on year, with a market share of 32.3%. Microsoft and Google ranked the second and the third in business revenue respectively. In 2019, Alibaba achieved a business revenue of $5.2 billion in cloud computing, up by 63.8% year on year, with market share rising to 4.9%. In regional distribution of Alibaba's

**Table 4.3** The distribution of global cloud computing market share in 2018–2019

| Cloud service provider | 2019 (×1 Billion USD) | Market share in 2019 (%) | 2018 (×1 Billion USD) | Market share in 2018 (%) | Growth rate (%) |
|---|---|---|---|---|---|
| Amazon cloud service | 34.6 | 32.3 | 25.4 | 32.7 | 36.0 |
| Microsoft cloud service | 18.1 | 16.9 | 11.0 | 14.2 | 63.9 |
| Google clound | 6.2 | 5.8 | 3.3 | 4.2 | 87.8 |
| Ali clound | 5.2 | 4.9 | 3.2 | 4.1 | 63.8 |
| Others | 43.0 | 40.1 | 34.9 | 44.8 | 23.3 |
| Total | 107.1 | 100.0 | 77.8 | 100.0 | 37.6 |

*Data Source* Canalys

business, in addition to Chinese market, Alibaba expanded rapidly in Southeast Asia and African countries. With strong coordination among regional business, it formed remarkable ecological pattern. See Table 4.3 for "The Distribution of Global Cloud Computing Market Share in 2018–2019".

3. Blockchain technology owns great potential in development and application

In 2019, 82 countries, regions and international organizations in the world published more than 600 blockchain-related policies. According to the statistics of 01 Finds, in 2012–2019, a total of 1,510 financing cases were completed in global blockchain field, with public financing amounting to ¥78.22 billion. Particularly, 2018 saw the "outburst" of blockchain financing. In 2019, financing in blockchain field was gradually rationalized, with the financing amount down by nearly 40% year on year (2018). In 2019, the number of financing deal in global blockchain field reached 543, with the financing amount of ¥23.83 billion. See Fig. 4.3 for "The Change & Trend of Amount and Number of Blockchain Financing in 2012–2019".

4. AI Triggers New Industrial Wave

AI integrates rapidly with various fields, boosts the transformation and upgrading of traditional industries, enhances the quality and efficiency of traditional industries, and stimulates new industrial wave globally. Using technologies like deep learning, global major technological giants plan the existing and future product systems in an all-round way, establish hardware and software product systems based on new-generation AI, deploy "software + hardware + application + chip" product pattern, and vie for user access and the dominant role in the new-generation AI industry. According to the basic statistics of Chinese Institute of Electronics, in 2019, global new-generation AI industry exceeded $70 billion in scale. Predictably, global AI industry scale would reach around $100 billion in 2020, with an average annual growth rate of 31.6% in 2018–2022.

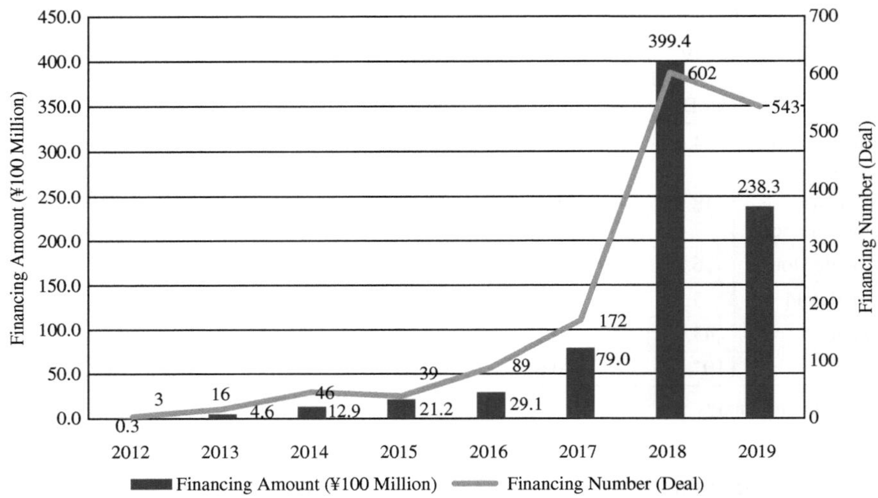


**Fig. 4.3** The change & trend of amount and number of blockchain financing in 2012–2019. (*Data Source* 01 Finds)

### *4.3.4 Internet Information Content Service Industry Expands Fast*

With continuous improvement and innovation in the Internet technology, information content service industry develops rapidly and Content Delivery Network (CDN) keeps growing, with great market potential. Pay-TV market declines quickly, and advertising video and streaming media win more support. Audiobook and Podcast continually grow. The United States takes the lead in audiobook market and China follows after. As 5G technology is fully developed and widely used, online game displays more potential.

1. Content delivery network continues to grow

Content Delivery Network (CDN) aims to improve the media quality, speed and reliability by physically connecting the content with user in a closer way. As Deloitte predicted, global CDN market would reach $14 billion in 2020, up by more than 25% over 2019. By 2025, CDN market will double to $30 billion, with a Compound Annual Growth Rate (CAGR) of more than 16%, as predicted.

2. Video advertising service industry starts to be transformed

In 2020, global video advertising service industry realized a revenue of $32 billion. Video advertising service industry in Asia reached $15.5 billion in revenue and accounted for nearly 50% of global revenue. The package service of streaming media cost much less than Pay-TV and provided consumers with unparalleled depth and diverse content. Video advertising market gradually turned from Pay-TV to streaming media. In the Asia–Pacific Region, stimulated by economical 4G networking service and low-cost smartphones, more than one billion Asian people watched ad-supported video. In the United States, media giants competed fiercely in streaming media market. Netflix, an American online video provider and producer, realized a business revenue of $5.77 billion in the first quarter of 2020 (FY), up by 27.6% year on year, with an increase of 15.8 million global paid subscribers. Netflix also realized a net profit of $709 million, up by 106.1% year on year.

3. New opportunities are opened to audiobook and podcast

As Deloitte predicted, global audiobook market scale would grow by 25% and reached $3.5 billion in 2020, and global Podcast market scale would grow by 30% and reached $1.1 billion (firstly exceeding $1 billion). The United States and China accounted for more than 70% of global audiobook market. In audiobook market scale, the United States ranked the first in the world and reached $1.5 billion in 2020, growing sustainably at an average annual growth rate of 20%-25% in next years. Besides, China would generate a revenue of $1 billion in 2020 in audiobook market, which was only $450 million in 2017.

4. The scale of online game users grows quickly

The explosive growth of mobile Internet, AR/VR and smart phone enables the number of mobile game player to continually rise. The Asia–Pacific Region will take up the largest market share in game industry. In 2021, the number of mobile online game player in Southeast Asia will increase to 250 million. After the outbreak of the COVID-19 Pandemic, the download of online game app increases fast, so do online time and online player. As the survey data in March 2020 suggested, in the first week of home quarantine, the number of video game player increased by 45% (ring growth) in the United States, by 38% in France, by 29% in the United Kingdom and by 20% in Germany. The number of online game users also jumps. 29% of American game players expressed that they spent much more time playing online games since the outbreak of the COVID-19 Pandemic. That was true to 17% of British game players and 12% of French game players.

## 4.4 Industrial Digitization Advances Deeply

The digital transformation of global economy quickens, and the digitization of manufacturing industry keeps deepening, which speed up the cross-border integration of industrial chain and feature with platform and shareability. The digital transformation and upgrading of service industry accelerates, and the digital transformation of agriculture develops steadily, on which various countries in the world concentrate in their development.

### *4.4.1 Agricultural Digitization Develops Firmly*

As a profound revolution, the digitization of agriculture and rural area covers that of agricultural element (like organism and environment), agricultural process (production, operation and management) and rural governance. Major developed countries in the world prioritize the development of digital agriculture and take it as strategic center. They have successively released strategies such as "Big Data Research and Development Plan", "Agricultural Technology Strategy" and "Agricultural Development 4.0 Framework" to forge new advantages in the new-round industrial revolution.

1. To promote the digital transformation of agriculture and rural areas is globally accepted

Globally, governments and organizations in major countries and regions successively launched "Digital Agricultural & Rural Development Plan". The United States built "AI Strategy Implementation Framework" and proposed "Smart Agriculture Research Plan". European Union issued *Horizon 2020*, which proposed to build smart service platform for small farmers by using earth observation technology. European Association for Agricultural Machinery presented "Agriculture 4.0" featured with the use of modern information technology and advanced agricultural machinery. Germany published *Organic Agriculture: Outlook Strategy*, which clearly aimed to develop smart agriculture based on the basic concept of "Industry 4.0". The Netherlands released *Digital Strategy*, which clarified the application of digital technologies into open farming, precision agriculture, greenhouse horticulture, animal husbandry and food quality & safety as well as the links of productive chain. Japan published *Japan's Robot Strategy* and started the "Next-Generation Innovative Technology for Agriculture, Forestry and Aquaculture Industries" based on "Intelligent Machinery + IT". South Korea released "Information Village Plan" to accelerate the informationization of rural areas, narrow the urban–rural gap and increase farmers' incomes.

2. Digital technology is widely used in agriculture

Digital technologies like big data, IoT, cloud computing, cognitive computing and AI integrate with agriculture in a closer way. Digital technology can be used to achieve precise control in agriculture. The United States provided farms with such services as planting-area measurement & calculation, crop-growth monitoring, growth-cycle estimation, yield prediction, natural-disaster prediction and early warning of diseases and pests, by collecting, processing and analyzing big data on meteorology via satellite as well as visualized system. In 2013, the United Kingdom formally started "Agro-Technology Strategy", which used digital technology, sensing technology and spatial geographic technology to develop planting and breeding industries more precisely, to construct a powerful platform for data collection, analysis and processing, and to strengthen the connection between agricultural production sector and market demand. Germany invested substantial fund to develop agricultural technology. In

April 2020, "Future Agricultural System: Digital Agricultural Knowledge and Information System" (DAKIS) was officially launched. It was funded by German Federal Ministry of Education and Research and co-developed by Leibniz Centre for Agricultural Landscape Research and other 9 research institutions. In five years, German government would provide a fund of 7.4 million euros and plan to develop a practical digital information and decision-support system, research robot, sensor and computer model, and combine productive optimization with the environmental and natural protection.

3. Technological giants successively deploy agricultural digitization

Agricultural digitization raises higher requirement on enterprise's technological level. World technological giants play an active role in the field of agricultural digitization. Now, Cisco develops an IoT-based technology that can remotely manage crops. It also invests into Prospera Technologies, a corporation that centers on computer vision and data processing, and develops an agriculture-dedicated AI system. Watson, a subsidiary of IBM, develops apps on precision agriculture and constructs digital models for agriculture via AI technology, which is mainly used to predict and simulate the uncertainty in future agriculture and to help farmers make more accurate decisions. Microsoft forges solutions for modern agriculture, by using machine learning, Cortana (an intelligent assistant) and IoT. In the seeding experiment in India, Microsoft uses AI technology to optimize the seeding process, which increases the average yield per hectare of farmland by 30%.

### *4.4.2 Manufacturing Digitization Continues to Deepen*

5G technology is gradually matured and widely used. This deepens the development of industrial Internet and builds the "ecosystem" of intelligent manufacturing. The wide use of industrial robot accelerates the digital transformation of manufacturing industry.

1. The scale of industrial internet market keeps rising

As the data published by CCID indicates, in 2019, global industrial Internet market scale grew by around 5% and reached $846.21 billion. Accenture predicts that by 2030, industrial Internet will produce an increment of more than $15 trillion of global economic volume. In the digitization of manufacturing industry, IoT is applied more widely, which helps enterprises to formulate strategies on productive procedure, cost and productivity. According to the research of MPI Group, about 70% manufacturers view that IoT raises their profitability. In 2020, manufacturing enterprises invested *c.* $267 billion into IoT. According to the data of U. S. National Association of Manufacturers, 90% American manufacturing enterprises employ less than 500 staff, owing to the use of digital technology like IoT.

2. 5G technology promotes the digital transformation of manufacturing industry

Connectivity has been the big problem that impedes the digital transformation of industrial manufacturing. 5G will play a great role in reducing delay and providing high bandwidth & large-scale reliable real-time communications. With 5G, manufacturers can enhance the use of sensor, cloud, quality testing and centralized tracking and build the "ecosystem" of intelligent manufacturing. In the future, the typical industrial use cases of 5G will extend from video monitoring, inspection & security and logistics & distribution (productive periphery) to product design & simulation, production control, quality testing and production safety (productive link), providing services for total factor and global value chain. For instance, Ericsson builds its first 5G smart factory in Texas, with an area of more than 2,780 $m^2$. It will use 5G to build the advanced antenna system to expand the network coverage.

3. Robot market scale continues to grow

In 2019, the global robot market scale reached $29.4 billion, with a growth rate of 3.2%. In specific, industrial robot market scale reached 15.9 billion US dollars, accounting for 54.1%. The Asia–Pacific Region owned the most active robot market, accounting for 60.2% of global share. Europe and North America followed, accounting for 19.9% and 17.4% respectively.

(1) The penetration rate of industrial robot continuously rises. In 2018, the sales of industrial robot increased by 5% (slightly down in 2019). In 2020, the sales of industrial robot would jump substantially by 10%, as predicted. The COVID-19 Pandemic further spurred the demand for industrial robots. As the report of Research and Markets suggested, in the post-COVID-19 era, global industrial robot scale (including peripheral device, software, and systematic engineering) would grow from $44.6 billion in 2020 to $73 billion in 2025, with a Compound Annual Growth Rate (CAGR) of 10.4%.

(2) Professional service in the robot market grows fast. As Research and Markets estimated, in 2020, around one million enterprise-edition robots would be sold globally. More than 50% of them were robots for professional service, including field robot, professional cleaning robot, medical robot and logistic robot, as well as robots used in defense, rescue and security. The robot market scale would exceed $16 billion, up by 30% over 2019.

### *4.4.3 Service Industrial Digitization Realizes Transformation and Upgrading*

As the data of Digital Market Outlook showed, in 2019, global major Internet service market reached around $183.61 billion in scale. The COVID-19 Pandemic promoted the innovation on business model of service industry and accelerated the development of Internet healthcare and online education.

1. Internet healthcare grows steadily

In the COVID-19 Pandemic, medical resources was vied for and medical supply was inadequate in some regions. This was represented by "three difficulties": difficulty in diagnosing the COVID-19 Pandemic, difficulty in seeking medical treatment for other patients and difficulty in distinguishing true and false information. Featured with no time–space limit, no contact and no cross-infection, Internet healthcare became a new way for citizens to seek medical treatment. As data suggested, global e-healthcare market would grow with a Compound Annual Growth Rate (CAGR) of 23% in 2019–2024.

2. Online education market presents explosive growth

As Internet becomes more popular and the new-type digital technology emerges, global online education market underwent explosive growth, impacted by the COVID-19 Pandemic. As Markets and Markets predicts, global online education market scale will grow from $8.4 billion in 2020 to $33.2 billion in 2025, with a Compound Annual Growth Rate (CAGR) of 31.4%. By 2027, global online education market scale will reach $54.05 billion. Regionally, North America will probably take up the largest market share, followed by Europe, the Asia–Pacific Region, the Middle East, Africa and Latin America. In particular, the demand for online learning will grow faster in the Asia–Pacific Region, including China, Japan and India.

3. The COVID-19 pandemic adversarially stimulates the rapid rise of teleworking

After the outbreak of the COVID-19 Pandemic, many enterprises arrange teleworking at home for their employees, in order to maintain normal social operation and reduce enterprise's loss. The public also express great concern over teleworking. In China, the penetration rate of teleworking remains relatively low; yet teleworking will probably develop quickly, as underlying technology and infrastructure like Internet, big data and cloud computing improve and traditional enterprise's demand for digital transformation becomes urgent. Since the COVID-19 Pandemic, Tencent Meeting, Ali DingTalk, CloudCC and other software have provided audio–video meeting, group live broadcasting and collaborative office services for tens of thousand small & medium-sized enterprises.

## 4.5 The Standardization of Fintech Parallels Its Development

Globally, Fintech thrives. Digital currency, digital bank and digital financial regulation become hot topics. Technological giants successively deploy Fintech field, with market competition increasingly fierce. Emerging market countries like Latin America become hot lands for Fintech development.

### *4.5.1 Issuing Digital Currency is Put on the Agenda*

Central banks of Ecuador, Senegal, the Republic of the Marshall Islands, Uruguay and Venezuela firstly launched digital currencies, in order to protect user's privacy, counteract money laundering, enhance the effect of monetary policy, strengthen national economic control capability and improve the competitiveness of national currencies. Central banks of China, Sweden, Bahamas, Eastern Caribbean, Lithuania, Thailand, Russia and Pakistan clearly proposed that issuing digital currency would be put on the agenda. Central banks of Canada, Brazil, Norway, the United Kingdom, Philippines, Israel, Denmark, Singapore and South Korea discuss and explore digital currency now.

### *4.5.2 Digital Bank Becomes a Popular Developmental Trend*

Digital bank develops prosperously. Emerging institutions form abundant resources and traditional banks deploy digital platforms. With sufficient financing, emerging digital banks develop vigorously. In the United States, Chime, a digital bank, raised $700 million, with its platform value exceeding $5.8 billion. In Latin America, Nubank, a Brazilian digital bank, obtained a one-off financing of $400 million. In the United Kingdom, Australia, Canada and Mexico, digital banks like Starling Bank, Tide, Xinjia Bank, Koho and Klar got the financing of more than 10 million US dollars. Meanwhile, traditional financial institutions successively deployed digital banks. HSBC Kinetic, HSBC's digital bank platform, received quick investment. Royal Bank of Scotland formally launched its digital bank product Bo. Singapore firstly opened digital bank license application access to guide the healthy industrial development. Malaysia released *Digital Bank Licensing Framework for Soliciting Comments*, which clarified the goal and plan for the development of digital bank.

### *4.5.3 Emerging Economies Show Vigor in Development*

In Latin America, Fintech market raised $2 billion. Brazil, Mexico, Colombia and Argentina attracted the most financing. Several venture capital firms like SoftBank and GGV Capital successively invested into the Fintech industries in Latin American market. Financial institutions that engaged in multiple fields like loan, insurance, digital bank, wealth management and mobile payment developed rapidly, e.g. Nubank (digital bank) and Creditas (digital loan platform). Mature Fintech platforms gradually turned to "one-stop service platform" model. Yet, as related technological and regulatory laws remained imperfect, Fintech markets in Chile, Peru and other countries or regions needed to be tapped.

### *4.5.4 The Supervision of Digital Finance Gradually Improves*

"Regulatory Sandbox" mechanism can provide moderately loose environment and larger space for operation and trial & error, on condition that regulation and supervision are ensured. So far, financial regulators in the United Kingdom, South Korea, Singapore, China and the United States have launched or will surely launch Fintech "Regulatory Sandbox". Simultaneously, the development of "Regulatory Sandbox" mechanism features with internationalization and differentiation. United States Securities and Exchange Commission, United States Office of the Comptroller of the Currency and United States Federal Deposit Insurance Corporation announced to join Global Financial Innovation Network. Monetary Authority of Singapore announced to launch "Fintech Sandbox Express Regulation Mechanism". British Financial Conduct Authority sought advice on "Cross-Sector Sandbox" and hoped to cooperate with several regulators in business coupling in controlled environment.

## 4.6 E-Commerce Keeps Expanding

Global e-commerce continues to expand. E-commerce sales hit new highs, and the proportion in global retail industry keeps jumping. Emerging markets like the Asia–Pacific Region grow quickly, and mobile e-commerce develops swiftly.

### *4.6.1 The Sales of E-Commerce Continually Rise*

The sales of e-commerce hit new highs, and the proportion in retail industry continually rises. As the data of eMarketer showed, in 2019, global e-commerce sales exceeded $3,466 billion and accounted for 13.7% of retail industry, up by 1.7% over 2018 (by 17.3% in 2021 as estimated). Noticeably, e-commerce plays an increasingly important role in global retail industry. See Fig. 4.4 for "Global E-commerce Sales and the Proportion in Retail Industry in 2017–2021".

(1) The growth rate of e-commerce sales slows down. Global e-commerce sales continue to rise, yet the growth rate presents a downward trend. In 2019, the growth rate of e-commerce sales reached 19.9%, down by 1.6% over 2018, despite the trend of high-speed growth. See Fig. 4.5 for "Global E-commerce Sales and the Growth Trend in 2017–2021".

(2) The proportion of mobile e-commerce increases stably. As smartphone and mobile payment become more popular, global mobile e-commerce sales display a trend of high-speed growth. In 2019, mobile e-commerce sales reached $2,224.08 billion

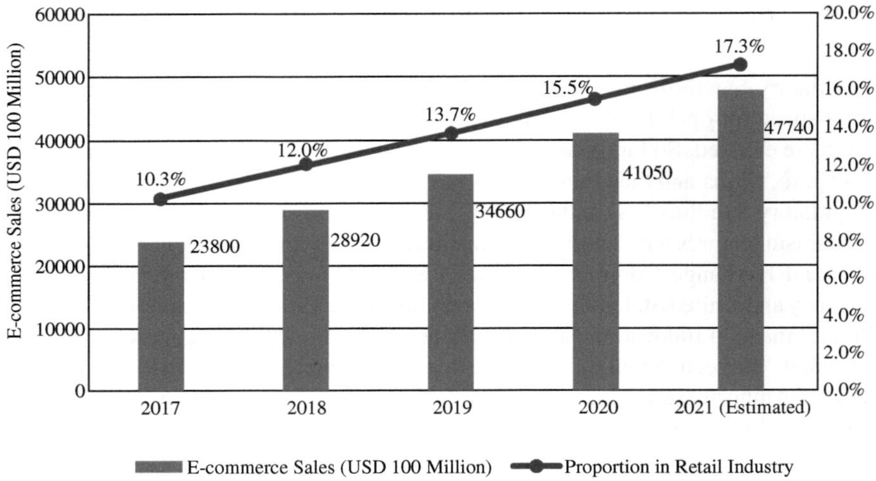


**Fig. 4.4** Global E-commerce Sales and the Proportion in Retail Industry in 2017–2021. (*Data Source* eMarketer)

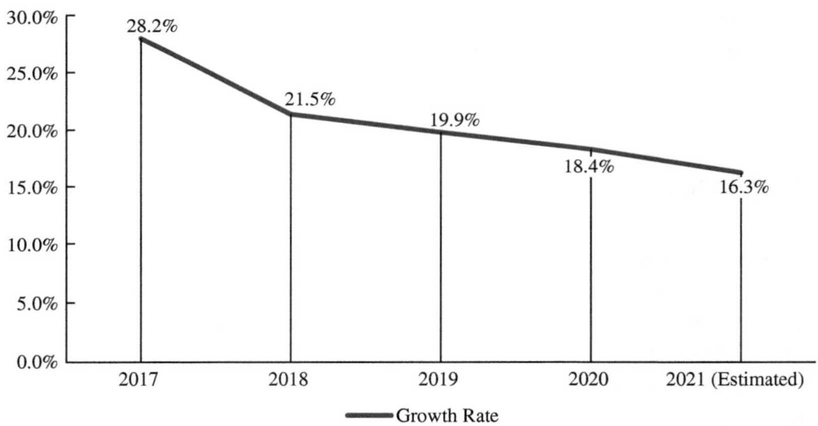


**Fig. 4.5** Global E-commerce sales and the growth trend in 2017–2021. (*Data Source* eMarketer)

and accounted for 64.2% of e-commerce sales, up by 3.4% over 2018. In 2021, the proportion will rise to 69.5%, as predicted. See Fig. 4.6 for "Global Mobile E-commerce Sales and the Proportion in 2017–2021".

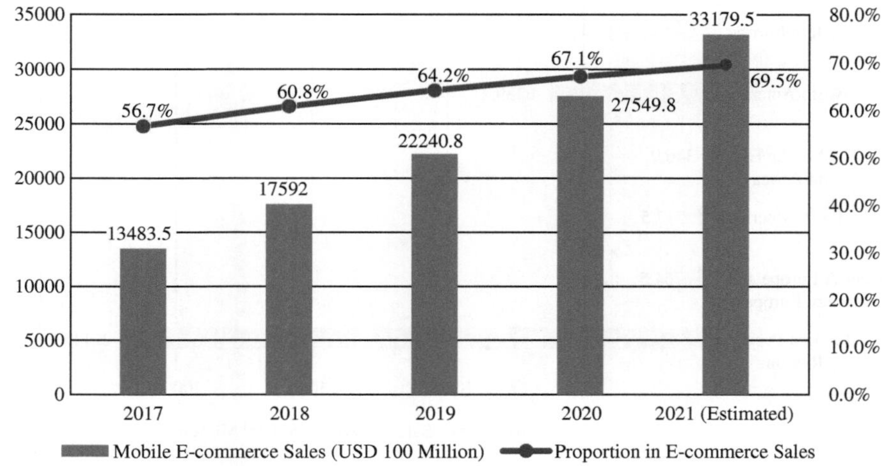


**Fig. 4.6** Global mobile E-commerce sales and the proportion in 2017–2021. (*Data Source* eMarketer)

### *4.6.2 Global Market Pattern Basically Remains Steady*

The Asia–Pacific Region maintains the dominant role in e-commerce transaction volume. In 2019, e-commerce transaction reached $2,211.42 billion in the Asia–Pacific Region, up by 23.9% year on year. Emerging e-commerce markets like Southeast Asia, the Middle East and India come to rise quickly. Take Vietnam as an example. As the data of Global Data suggested, transaction volume in its e-commerce market doubled in past five years, increasing from 90.1 trillion VND (approximately $3.9 billion) in 2015 to 218.3 trillion VND (approximately $9.4 billion) in 2019. According to the statistics of Payments Bureau of State Bank of Vietnam, in 2019, the transaction volume of Internet payment in Vietnam rose by 66%. See Fig. 4.7 for "Global E-commerce Transaction Volume in Various Regions in 2019".

In the growth rate of market transaction volume, in top ten countries with the fastest growth in e-commerce in 2019 (see Fig. 4.8), six come from the Asia–Pacific Region. The growth rate of e-commerce transaction volume exceeded 30% in India and the Philippines, and reached 25.9% in Chinese market, Malaysia (22.4%), Indonesia (22%), Singapore (21.5%) and South Korea (17.5%) followed them.

In North America and Western Europe, the growth rate of e-commerce transaction volume stays lower than global average level. In 2019, the growth rate of e-commerce transaction volume in North America reached 13.2%, and that in Western Europe reached 10.6%, both lower than global average level (19.9%). In the Middle East, Africa and Latin America, e-commerce transaction volume grow faster. As relevant data showed, in 2019, e-commerce transaction volume rose by 22% in Latin America, by 21.3% in the Middle East and Africa, both higher than global average level. See Fig. 4.9 for "Growth Rate of E-commerce Transaction Volume in Various Regions in 2019".

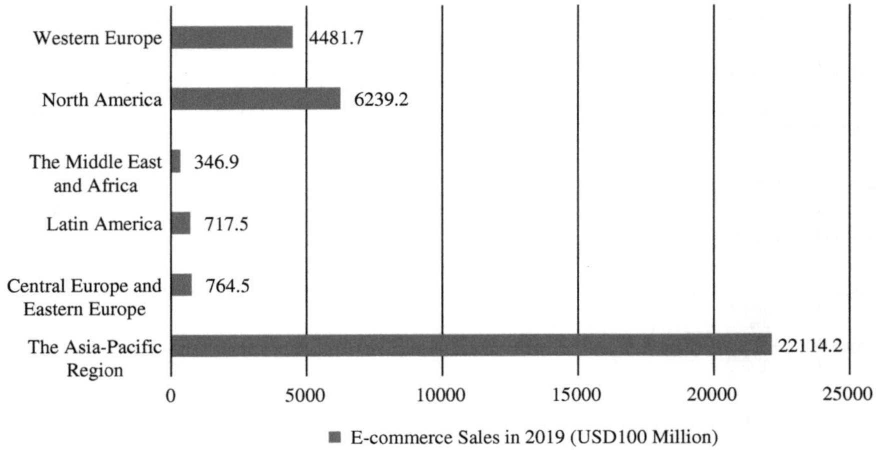


**Fig. 4.7** Global E-commerce transaction volume in various regions in 2019. (*Data Source* eMarketer)

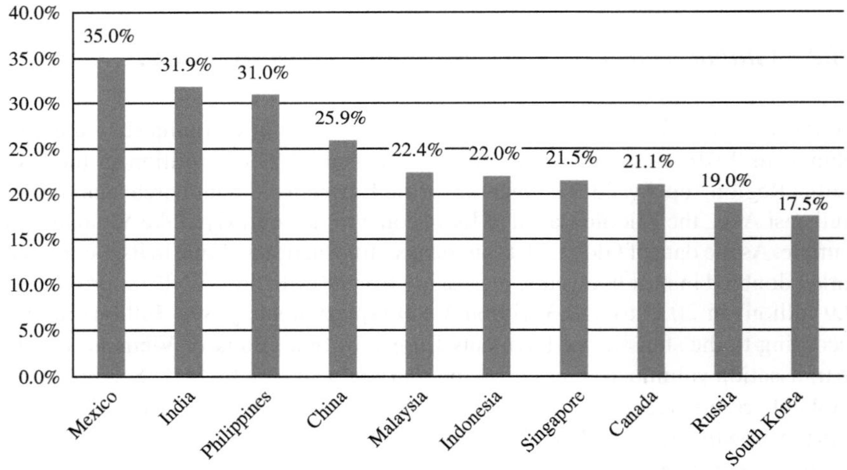


**Fig. 4.8** Top ten countries with the fastest growth in E-commerce in 2019. (*Data Source* eMarketer)

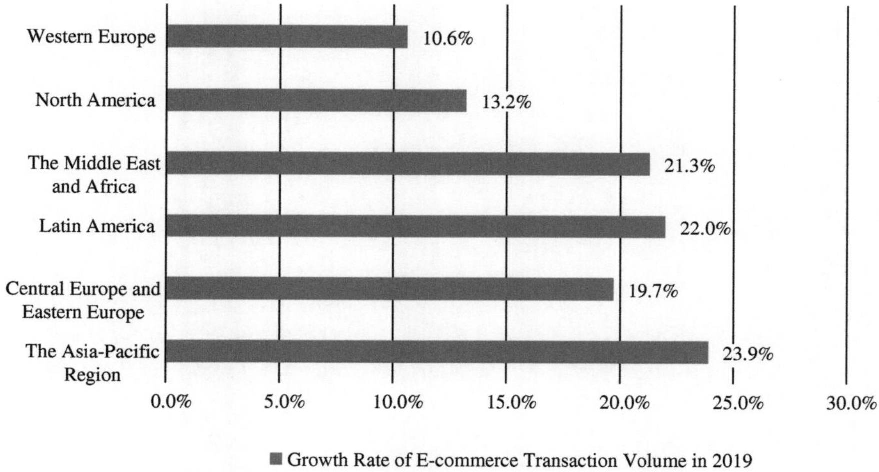


**Fig. 4.9** Growth rate of e-commerce transaction volume in various regions in 2019. (*Data Source* eMarketer)

# Chapter 5
# The Development of World E-Government

## 5.1 Outline

In recent years, information and economic-social field integrate quickly, and various countries widely explore the use of information in governmental affairs. In technological application, the application and innovation of new-generation information technologies like blockchain, IoT and AI continue to emerge. In service supply, various countries concentrate on actively providing governmental service, positively guiding public value and effectively making governmental response, with service channel and content continuously expanded. In governance model, innovation is continually made in collaborative governance, and some management methods and governmental procedures are re-established and re-formed. As international major reports on e-government development suggest, countries in the world develop well in governmental infrastructure, governmental data openness, online governmental service and digital literacy. In e-government, Denmark, Sweden, the United States and Singapore take the lead, and China achieves remarkable progress.

E-government also plays a positive role in counteracting the COVID-19 in various countries. Facing the spread of the COVID-19, World Health Organization (WHO) actively uses e-government and guides and helps countries in the world to effectively combat the COVID-19. Singapore, Germany, Italy and South Korea successively adopt e-government to enhance social management and prevent & control the COVID-19.

## 5.2 The Leading Exploration of World E-Government

In recent years, in world e-government, a series of innovations and breakthroughs are achieved in technological application, service model and governance collaboration, with rich experience accumulated, theoretically and practically.



Chinese Academy of Cyberspace Studies, *World Internet Development Report 2020*,
https://doi.org/10.1007/978-981-16-9388-5_5

### 5.2.1 The Innovation on Technology and Application Emerges in E-Government

1. Blockchain Becomes a Hot Topic in Governmental Affairs

Blockchain boasts prominent advantages in data security & management.[1] As blockchain technology gradually matures, various countries constantly improve blockchain-related laws and regulations, propose strategic initiatives, and make innovations in fields like data asset and security management. The European Union launched *Horizon 2020*, which planned to invest 300 million euros to support the development of cross-border blockchain ecosystem in Eurozone.[2] In 2019, German government released *German Federal Government Blockchain Strategy*, which aimed to tap the potential of the digital transformation of government and provide digital public administrative service by using blockchain technology.[3] Some countries or regions clarified the role of blockchain in legal terms. State of Arizona (the United States), for example, enacted the law that clearly conferred equal legal effect on blockchain intelligent contracts with common contracts.[4]

Presently, blockchain technology is mainly applied to cases with high demand for security, like finance, taxation and identity authentication. European countries, including Norway, Malta, Spain, Greece and the United Kingdom, jointly forge "Use Case Diplomas" Project, which facilitates the exchange of diplomas among member states by establishing European Trust Network. Dutch government explores the application of blockchain technology to various governmental businesses.[5] Centraal Justitieel Incassobureau (CJIB) launches Financial Emergency Stop Project to create a financial information security ecosystem, in which citizens can access personal debt data and choose trusted organizations before sharing their data. Dutch tax authority explores the application of blockchain technology to business tax. In counteracting the COVID-19, Dutch government also uses blockchain technology to ensure the transparency and information security of supply–demand docking in medical products. Dutch Red Cross website also accepted donations paid in Bitcoin.[6]

2. Various Emerging Technologies Penetrate into Public Fields Quickly

Technological application and innovation constitute an important factor that drives the development of e-government. New-generation information technologies like IoT, AI and cloud computing enormously reshaped governmental governance in the last decade. This ensured the supply of governmental service, improved the

[1] Liu and Yang (2020).

[2] https://www.trustnodes.com/2018/04/11/european-blockchain-partnership-signed-e300-million-allocated-blockchain-projects.

[3] https://www.bmwi.de/Redaktion/EN/Publikationen/Digitale-Welt/blockchain-strategy.html.

[4] https://www.azleg.gov/legtext/53leg/1r/bills/hb2417p.pdf.

[5] Koster and Borgman (2020).

[6] Dutch Government to use Blockchain to fight Covid19 https://www.unlock-bc.com/news/2020-03-31/dutch-government-to-use-blockchain-to-fight-covid19.

transparency of governmental information, and altered the way of government-citizen interaction. Besides, the reform of governmental organizations and the optimization of governmental procedures were continuously promoted.

IoT technology is widely used in such fields as smart city, smart community and smart home. For example, Santander, a Spanish city, launches "Smart Santander" Platform,[7] which deploys sensing devices in urban facilities like street lamps, bus stations, buildings, buses and taxis, and covers smartphones installed with relevant Apps in the platform (as important sensing devices). With these digital devices, real-time monitoring of urban operation can be achieved. In environmental monitoring, Santander installs more than 2,000 IoT devices on lampposts and buildings' surfaces in the center to measure different environmental parameters like temperature, carbon monoxide, noise and light. To irrigate city parks more efficiently, Santander deploys around 50 devices in two key green zones to monitor irrigation-related parameters like humidity and wind condition.

Machine learning technology can automatically classify and process multi-source heterogeneous data.[8] For example, Audrey, an AI system of the National Aeronautics and Space Administration (NASA), can assist emergency managers in cognitively reasoning and making decisions and realizing immediate response under emergency.[9,10] The system has been piloted in Grant County, Washington, and Ontario, Canada.[11] London (UK) cooperates with Alan Turing Institute and analyzes open environmental data via machine learning technology to reduce air pollution.[12]

### 5.2.2 E-Government Stimulates Innovation in Service Model

1. Governments Proactively Provide Governmental Services

Traditionally, governmental service model is passive. By contrast, the service model driven by e-government attaches more attention to public-value-oriented proactive service. Estonia boasts the country that achieves digital transformation fastest in the world. In the *United Nations E-Government Survey*, Estonia's ranking rises from the 16th in 2018 to the third in 2020. "Proactive Services" is an important way for the its digitalized development.[13]

[7] Sánchez et al. (2020).

[8] Usman et al. (2019).

[9] https://gcn.com/articles/2019/09/25/ai-audrey-responders.aspx.

[10] https://technology.nasa.gov/features/audrey.html.

[11] https://healthcare-in-europe.com/en/news/could-ai-audrey-be-the-future-first-response-assistant.html.

[12] https://mobility.here.com/learn/smart-city-initiatives/london-smart-city-tackling-challenges-20-initiatives.

[13] https://e-estonia.com/estonia-top-3-in-un-e-government-survey-2020/.

**Fig. 5.1** Estonian government propagates the first governmental proactive service on Facebook: family benefits service starts[14]

Estonian government proactively provides "seamless" service for citizens.[15] By technological means, various governmental departments can connect their services that cover major stages of citizens' life (birth, schooling, job hunting and house purchase) and realize the pattern "One Institution Serves Citizens in an All-Round Way". Take Estonia's first event-based "family benefits service" (see Fig. 5.1) as an example. It is a welfare for all Estonian citizens in charge of the Estonia Social Insurance Board (SKA). In its running, citizens needn't apply for it, as Estonian government proactively provides citizens with corresponding benefits based on the information it owns.[16] For example, after the birth of a baby, SKA sends a mail to the baby' parents on getting relevant family benefits. Now, the service provides Estonian citizens with monthly family benefits of more than 44 million euros.

2. Governments Improve Citizens' Effective Participation

Various countries continuously promote citizens' participation in public affairs via e-government. For instance, Madrid (Spain) council establishes "Decide Madrid", an online public participation website platform,[17] to promote citizens' participation in governmental decisions. Users can initiate proposals and vote on the platform. "Decide Madrid" boosts citizens' effective participation in public affairs while ensuring the transparency of municipal procedures in Madrid.

---

14 https://www.facebook.com/estoniadigitalsociety/posts/the-first-proactive-governance-service-is-now-live-from-now-on-new-parents-need-/3093571767383980/.

15 https://e-estonia.com/proactive-services-estonia/.

16 https://news.err.ee/991789/parents-no-longer-have-to-apply-for-family-benefits.

17 https://decide.madrid.es/.

### 5.2.3 E-Government Triggers Coordinated Innovation in Governance

1. Intra-Governmental Coordination

In recent years, the level of intra-governmental coordination continuously improves and progresses toward whole-process & all-round governmental service. In many countries, governments coordinate data standards, reconstruct business procedures, and realize data exchange and business collaboration among various governmental departments, by building integrated governmental platforms. For example, Uruguayan government develops the Uruguayan e-Government Interoperability Platform (egovIP), which forms the foundation for data interoperability among institutions.[18,19] Through egovIP, various systems and devices can exchange data from diverse sources in a standardized way. In *Digital Government Strategy 2020*, Uruguayan government proposes to further develop data interoperability plans, e.g. opening up national spatial data and setting up data standards.

2. Multi-Subject Coordination

The multi-subject coordination of governments, enterprises and citizens displays an important development trend for the supply of public service products. The collaborative model helps meet the demand for public service, improve the transparency of government, ease financial pressure and promote the innovation in public service.

In recent years, some countries build "collaborative ecology" driven by digital platforms. Governments provide digital environment for innovation & coordination according to public needs rather than specific services (e.g. approval, licensing and infrastructure). For instance, "My Renovation Planner", an online government-enterprise collaboration platform in New South Wales (Australia), forges innovation & application environment for urban citizens and designers, which meets their customized demands for decoration and facilitates governmental monitoring & supervision of market transactions.[20] In the Netherlands, Port of Rotterdam, ABN AMRO Bank and Samsung have jointly developed Deliver, a digital platform, and created a collaborative digital environment for sharable logistics and financial information.

Presently, collaborative governance stresses the early participation of stakeholders to reduce the collaborative risk at the later stage. In order to determine the needs of all parties as early as possible and make them cooperate & coordinate in advance, Ministry of Finance of Finland initiates "Lupapiste.fi" Project in *Electronic Service and Electronic Democracy Action Plan.* At an early stage, it invites stakeholders to participate in and responds to relevant needs in a timely manner, which effectively raises the level of government-citizen interaction.[21]

---

[18] González and Ruggia (2020).

[19] González et al. (2016).

[20] Tan et al. (2020).

[21] Helander et al. (2020).

## 5.3 Practical Evaluation of World E-Government

In recent years, international organizations and research institutions, including the United Nations (UN), World Information Technology and Services Alliance (WITSA) and Waseda University, continually evaluate e-government, digital government and digital competitiveness, and summarize & assess e-governments in various countries and regions from multiple dimensions (e.g. comprehensive development, governmental infrastructure, governmental data openness, online service level and digital literacy level). This enables the public and governments to understand the condition of world e-government development.

### *5.3.1 Comprehensive Evaluation of E-Government*

Currently, international comprehensive evaluative reports on the development of e-government mainly include: the *United Nations E-Government Survey* released by the United Nations,[22] *World Digital Competitiveness Ranking* released by Institute for Management Development (IMD, Lausanne, Switzerland), *Networked Readiness Index* released by World Information Technology and Services Alliance (WITSA),[23] and *International Digital Government Rankings Evaluation Report* released by Waseda University. These evaluative reports center on different topics. UN's report measures the e-government development in a country from such aspects as Online Service Index (OSI), Telecom Infrastructure Index (TII) and Human Capital Index (HCI). IMD's report concerns knowledge, technology and readiness for the future, and evaluates the capacity and readiness in different economies that use digital technology as main driving force for economic, governmental and social transformation. WITSA's report evaluates the readiness of e-government network from four aspects, i.e. technology, manpower, governance and influence.[24] Waseda University uses 10 indicators (like network readiness and online service) to measure world e-government development. See Table 5.1 for "Recent Ranking & Comparison of Some Countries in Four Comprehensive E-government Evaluation Reports".

Generally speaking, Denmark, Sweden and the United States maintain a relatively high level in digital construction and e-government development, and rank top 10 in the world in various evaluative reports. China remains above the average level and rises stably. In the *United Nations E-Government Survey* released by the United Nations, China's overall ranking leaped from the 65th in 2018 to the 45th in 2020.

---

[22] https://publicadministration.un.org/en/research/un-e-government-surveys.

[23] https://networkreadinessindex.org/.

[24] https://networkreadinessindex.org/.

**Table 5.1** Recent ranking & comparison of some countries in four comprehensive E-Government evaluation reports

| | UN 2020 | IMD 2019 | WITSA 2019 | Waseda 2018 |
|---|---|---|---|---|
| Countries Covered | 193 | 63 | 121 | 65 |
| Denmark | 1 | 4 | 6 | 1 |
| South Korea | 2 | 10 | 17 | 6 |
| Estonia | 3 | 29 | 23 | 4 |
| Finland | 4 | 7 | 7 | 13 |
| Australia | 5 | 14 | 13 | 10 |
| Sweden | 6 | 3 | 1 | 8 |
| The United Kingdom | 7 | 15 | 10 | 3 |
| New Zealand | 8 | 18 | 16 | 14 |
| The United States | 9 | 1 | 8 | 5 |
| The Netherlands | 10 | 6 | 3 | 17 |
| Singapore | 11 | 2 | 2 | 2 |
| Iceland | 12 | 27 | 21 | 15 |
| Norway | 13 | 9 | 4 | 11 |
| Japan | 14 | 23 | 12 | 7 |
| Austria | 15 | 20 | 15 | 22 |
| China | 45 | 22 | 41 | 32 |

*Note* Data in Table 5.1 sorted out by the author

### *5.3.2 Infrastructure of E-Government*

Technology indicator and network readiness indicator in IMD's *World Digital Competitiveness Ranking*, as well as technology indicator in WITSA's *Networked Readiness Index*, represent the construction of e-government infrastructure in various countries to some degree.

IMD's evaluation consists of two aspects: technology indicator and network readiness indicator. In technology indicator, regulatory framework refers to the integrity of the regulatory frameworks for establishing business, enforcing contract and legislation on research & intellectual property rights in various countries. Capital indicator indicates the condition of venture capital and telecom investment in various countries. Technological framework includes such indicators as communication technology, mobile broadband users, wireless broadband, number of Internet users, Internet bandwidth speed and high-tech export. In network readiness indicator, the adoption attitude denotes the application of electronic participation and smartphones. Business agility includes big-data analysis-tool application, knowledge transformation and robot application & distribution. IT integration reflects e-government, public–private partnership (PPP), cybersecurity and pirated-software management. See Table 5.2

**Table 5.2** Comparison of various indicators between top 15 countries and regions in IMD's evaluative list and Chinese mainland

| Countries | Rank | Technology indicator | | | Network readiness indicator | | |
|---|---|---|---|---|---|---|---|
| | | Regulatory framework | Capital | Technological framework | Adoption attitude | Business agility | IT integration |
| United States | 1 | 19 | 1 | 11 | 2 | 2 | 5 |
| Singapore | 2 | 2 | 8 | 1 | 19 | 6 | 4 |
| Sweden | 3 | 5 | 4 | 12 | 8 | 13 | 12 |
| Denmark | 4 | 10 | 27 | 8 | 1 | 10 | 1 |
| Switzerland | 5 | 14 | 16 | 9 | 11 | 14 | 7 |
| The Netherlands | 6 | 6 | 5 | 10 | 9 | 7 | 3 |
| Finland | 7 | 9 | 11 | 13 | 6 | 27 | 2 |
| Hong Kong (China) | 8 | 12 | 6 | 3 | 12 | 8 | 22 |
| Norway | 9 | 3 | 7 | 6 | 5 | 23 | 9 |
| South Korea | 10 | 26 | 29 | 7 | 4 | 5 | 21 |
| Canada | 11 | 17 | 10 | 27 | 17 | 16 | 13 |
| United Arab Emirates | 12 | 1 | 2 | 5 | 20 | 4 | 8 |
| Taiwan (China) | 13 | 23 | 12 | 4 | 14 | 3 | 24 |
| Australia | 14 | 7 | 19 | 17 | 7 | 35 | 11 |
| The United Kingdom | 15 | 18 | 22 | 18 | 10 | 26 | 14 |
| Chinese Mainland | 22 | 20 | 32 | 32 | 24 | 1 | 41 |

for "Comparison of Various Indicators between Top 15 Countries and Regions in IMD's Evaluative List and Chinese Mainland".

The United States ranks the first in capital indicator and top 5 in network readiness indicator. Singapore takes the lead in the world in regulatory framework, technological framework and IT integration, but ranks low in adoption attitude. Chinese Mainland ranks the first in business agility indicators, yet lags behind in IT integration, capital, technological framework and other indicators.

Remarkably, in 3-year IMD's ranking of indicators, e.g. world digital competitiveness technology (see Fig. 5.2) and network readiness (see Fig. 5.3), changes happen. The United Arab Emirates, the Netherlands, France, Latvia, Lithuania, Chinese Mainland, Thailand and Spain keep advancing in technology indicator. South Korea, Ireland, China (including Hong Kong and Taiwan), Germany and Luxembourg make great progress in network readiness.

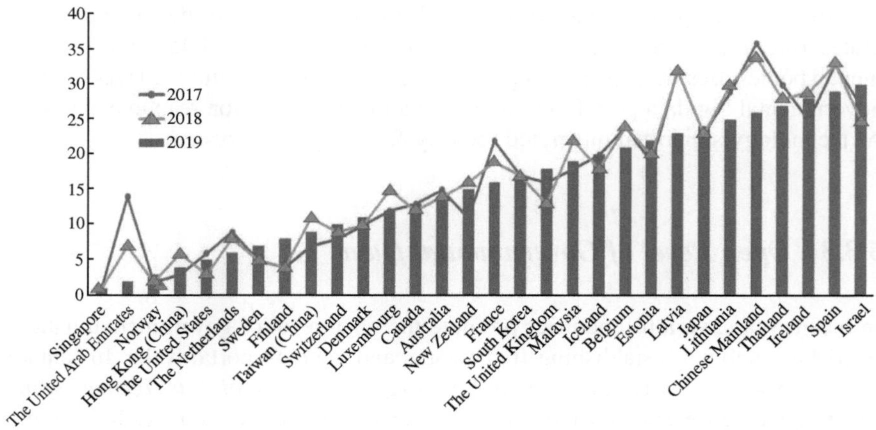


**Fig. 5.2** IMD's ranking of world digital competitiveness technology (Top 30)

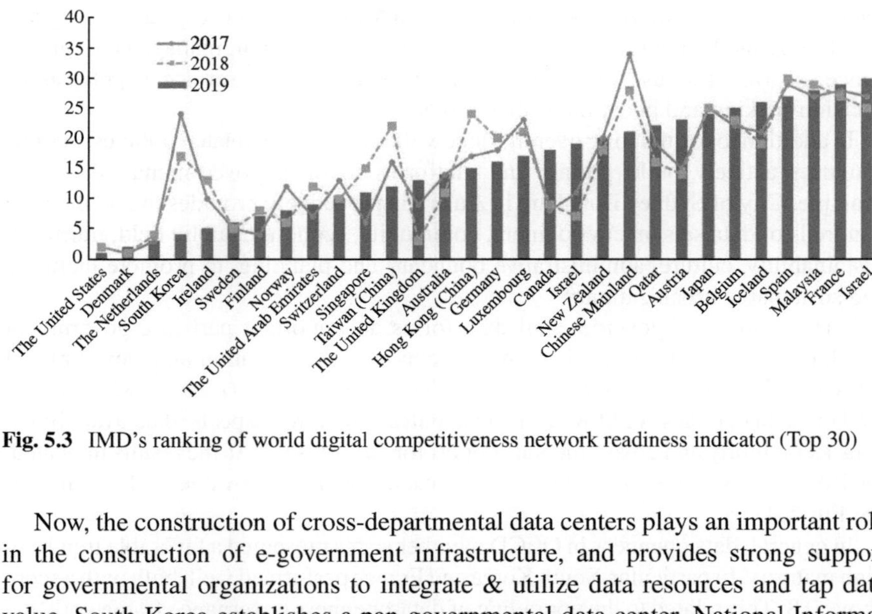


**Fig. 5.3** IMD's ranking of world digital competitiveness network readiness indicator (Top 30)

Now, the construction of cross-departmental data centers plays an important role in the construction of e-government infrastructure, and provides strong support for governmental organizations to integrate & utilize data resources and tap data value. South Korea establishes a pan-governmental data center, National Information Resources Service (NIRS), which integrates and manages data and information of central governmental bodies. NIRS promotes the sharing of information among departments and optimizes the utilization of resources via government cloud. It integrates, runs and manages 1,230 types of e-government service of 45 central governmental bodies. Besides, it supports governmental decisions by running Hye-An (pan-governmental big data portal) and protects national information resources by using AI technology-supported integrated security & management systems.[25]

### *5.3.3 Open Level of Governmental Data*

In recent years, various countries promote the open level of governmental data by publishing policies, establishing frameworks and opening portal webs. In January 2020, Norway released the *National Strategy for Artificial Intelligence*, which stressed that government should consolidate the foundation for AI by opening data and promote the innovative use of data among academia, businesses and society. 9 countries including Estonia, New Zealand, South Korea and the United Kingdom initiated and organized Digital 9 (D9), a digital states alliance. In 2019, D9 released *Data 360° Montevideo Declaration* in Montevideo, the capital of Uruguay, which presented a common vision on the creation, collection, management, governance, sharing and use of governmental data to improve service supply, assist decision-making and boost innovative research.

In addition to formulating overall frameworks and issuing related policies, various countries actively build portals and platforms to open governmental data. For example, City of Sydney Data Hub in Australia proactively provides the public with hundreds of data sets on environment, community, economy, public field, transport, sustainability, culture, administrative boundary and planning, to promote inclusive and sustainable urbanization.[26]

The openness of governmental data forms an important part in e-government development. In 2019, Organization for Economic Co-operation and Development (OECD) released *Open, Useful and Re-usable data* (OURdata) *Index*,[27] which evaluated the data openness of OECD member states from three aspects: data availability, data accessibility and governmental support for data reuse.[28] As the results indicated, OURdata Index of OECD member states reached an average value of 0.6 (out of 1, see Fig. 5.4).

In general, data openness in OECD member states presented a favorable trend and high maturity. In particular, South Korea and France performed well, both with a score above 0.8. This related to the upgrading of data-openness governance framework, the rise of political will, the improvement of data-reuse environment and the continuous advancement of national high-level in South Korea and France. In Open Government

[25] Source: South Korea National Information Resources Service (www.nirs.go.kr).

[26] City of Sydney, Data Hub (https://data.cityofsydney.nsw.gov.au/).

[27] http://www.oecd.org/gov/digital-government/open-government-data.htm.

[28] The index was released in 2015 and is updated every two years.

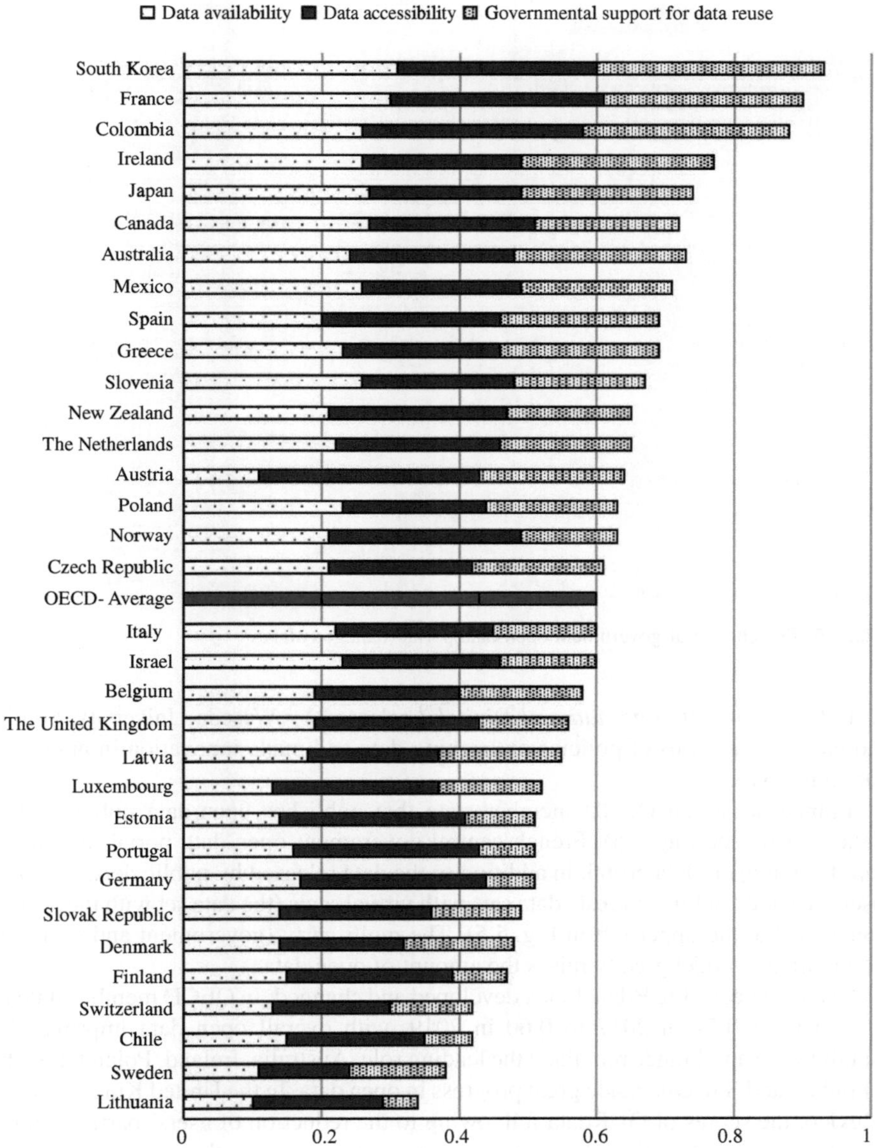


**Fig. 5.4** Governmental data index published by OECD

**Fig. 5.5** French central-government open-data portal (data.gouv.fr)

Data (OGD) in *IAC International Digital*[29] released by Waseda University, South Korea issued a series of policies on opening data and made innovation in practice, ranking top too.

France boasts an OECD member state that publishes open-data policy early. data.gouv.fr (see Fig. 5.5), French central-government open-data portal, develops rapidly. On open-data portal, in addition to the data released by public departments, users can add "public interest" data sets with virtual seals (the data set with the √sign round seal at the upper left in Fig. 5.5). The multi-party (government and society) collaborative model greatly raises the amount of open data.

In recent years, OURdata Index developed and changed. In OECD member states, it rose from 0.54 in 2017 to 0.60 in 2019, with overall open data improved.[30] South Korea and France remained the leading role. Australia, Ireland, Poland, Czech Republic and Slovenia made great progress in open data. In the United Kingdom and Mexico, the scores of OURdata fell, owing to the reduction of users' participation, the shift of central work and the change of regulatory organization.

---

[29] https://iacio.org/wasada-iac-world-e-government-ranking/.

[30] http://www.oecd.org/governance/digital-government/ourdata-index-policy-paper-2020.pdf.

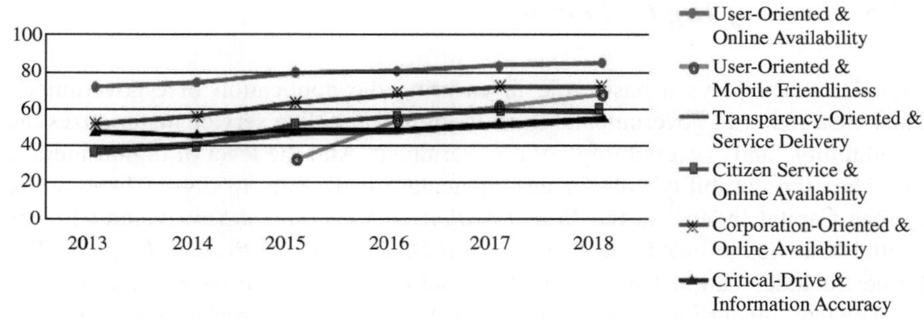


**Fig. 5.6** Change of online service index in European countries in 2013–2018

## 5.3.4 Level of Online Service

Presently, major reports that evaluate the level of online service include the *United Nations E-Government Survey* and *EU eGovernment Benchmark Report*. *EU eGovernment Benchmark Report*[31] examines user-centered online governmental service provided in 28 EU member states from three aspects, i.e. the accessibility, mobile friendliness and availability of online service. In 2019, as the evaluative result suggested, the average score in accessibility of online service reached 85. Malta, Portugal and Estonia ranked the first, the second and the third respectively, all above 98. This indicated that governmental services were highly digitalized in these countries. In the COVID-19, in Estonia, 99% of governmental service could be completed online[32] and nearly 70% of governmental service be completed in mobile devices.

In 2020, the evaluative result of UN Online Service Index (OSI) suggested that European countries dominated "very high" and "high" OSI groups. *EU eGovernment Benchmark Report* also represented overall highly developed level of e-government service in European society. See Fig. 5.6 for "Change of Online Service Index in European Countries in 2013–2018". Generally speaking, the average level of online service rose stably in Europe. "User-Oriented & Online Availability" index jumped from 72% in 2013 to 85% in 2018. "User-Oriented & Mobile-Friendliness" index improved quickly, from 33% in the first evaluation in 2015 to 68% in 2018. "Critical-Drive & Information Accuracy" was used to measure the accuracy of information pre-filled by public institutions. Malta, Estonia and Lithuania scored the highest in the index, all above 88%.

[31] https://www.capgemini.com/resources/egovernment-benchmark-2019/.

[32] https://e-estonia.com/digital-society-during-covid-19-lock-down/.

### 5.3.5 Level of Digital Literacy

Digital literacy plays a basic role in ensuring the application of e-government. The construction of governmental infrastructure and online service materializes the "availability" and "accessibility" of e-government. And the level of digital literacy embodies the "usability" of citizens. Indicators that relate to digital literacy are Human Capital Index[33] in the *United Nations E-Government Survey* and Citizens Digital Knowledge Index[34] in IMD's *World Digital Competitiveness Ranking.* The former includes national literacy rate, school enrollment rate, years of education, etc. The latter underlines education, science & technology, experience, training, etc. See Fig. 5.7 for "Ranking of Second-Level Index on Digital Knowledge in IMD's *World Digital Competitiveness Ranking* in 2019". In general, Europe, the Americas and East Asia maintain relatively high levels in digital knowledge, and Chinese Mainland remains above the average level.

According to the ranking of second-level index on digital knowledge (top 30 countries and regions) in IMD's *World Digital Competitiveness Ranking* in 2017–2019 (see Fig. 5.8), the United States, Switzerland, Sweden, Denmark and Austria rank top 10 and rise stably. Singapore drops to the third in the world, from the first for three consecutive years. Besides, Canada and Israel fall slightly in the ranking. Compared with 2018, Asian economies jump remarkably in the ranking. China's Hong Kong and South Korea rank top 10, and Chinese Mainland ranks the 22nd. Supported by positive achievements made in talent and training & education, India and Indonesia also climb four and six places respectively in the ranking.

E-government reports above-mentioned commonly view that digital skills help to achieve social inclusiveness, which ought to be enhanced among civil servants and public departments. In recent years, Bangladesh strives to raise the digital skills of governmental officials, promote data-based development planning, public-service design and policy implementation, and accelerate Bangladesh's extrication from the least developed countries list by 2024. In 2019, with the support of the United Nations Department of Economic and Social Affairs, Bangladesh collaborated with National University of Singapore and Temasek Foundation on providing training for Bangladeshi officials, in order to enhance civil servants' ability in data collection, use and processing. In recent years, Kazakhstan government strengthens the abilities of chief digital officers and IT specialists of various-level governments, trains civil servants in ICT skills and boosts the digital transformation of governments, via "Digital Kazakhstan" Strategy.[35,36]

---

33 https://m.jicmian.com/article/4673815.html.

34 https://www.imd.org/wcc/world-compctitiveness-ccnter-rankingsworld-dinital-competitiven cess-ranking,2019.

35 https://publicadministration.un.org/en/UNPSA.

36 https://digitalkz.kz/en/about-the-program.

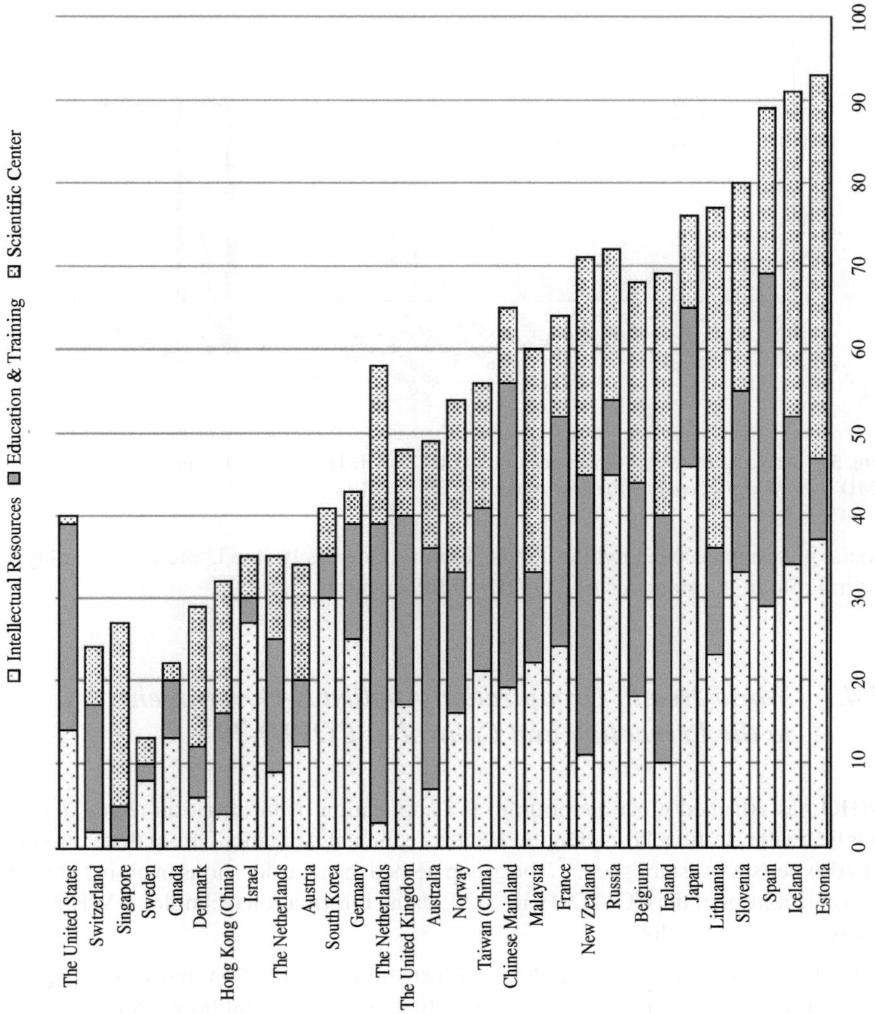


**Fig. 5.7** Ranking of second-level index on digital knowledge in IMD's *world digital competitiveness ranking* in 2019

## 5.4 Annual Hot Words: E-Government Under the Outbreak of COVID-19

The COVID-19 broke out in early 2020 and impacted tremendously on economy and society of various countries in the world. International organizations and governments, including World Health Organization (WHO), Germany, Italy and Singapore, used information technology to curb the spread of the COVID-19 and implement

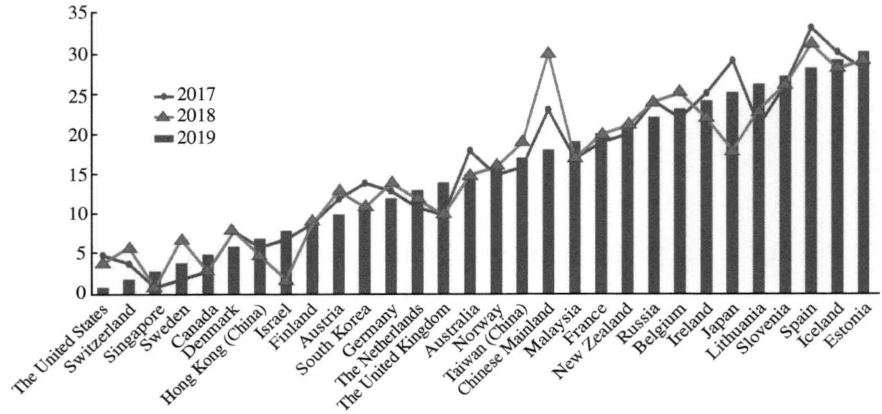


**Fig. 5.8** Ranking of second-level index on digital knowledge in top 30 countries and regions in IMD's *world digital competitiveness ranking* in 2017–2019

social governance. Meanwhile, Johns Hopkins University (the United States) played a vital role in the statistics of COVID-19 information.

## *5.4.1 World Health Organization Applied E-Government to the Prevention and Control of COVID-19*

WHO has attached great attention to the application of information technology to public health. In June 2019, Digital Health Technical Advisory Group that consisted of 20 experts was established.[37] The group helped to make decisions in global prevention and control of the COVID-19 in 2020. Using information technology, WHO took measurements as follows.

(1) Developing robots on epidemic-information consultation and promoting the release and propagation of information on the prevention & control of the COVID-19. On 20 March 2020, WHO worked with WhatsApp, a mobile app provider. They launched WHO Health Alert chatterbot that worked in seven languages, i.e. Arabic, English, French, Hindi, Italian, Spanish and Portuguese. Users could obtain the latest information on the COVID-19 by starting conversational connection on WhatsApp. In the end of March 2020, WHO announced to cooperate with Rakuten Viber[38] and launch a chatterbot, by which users would directly receive the notification on the latest news and information from WHO, after subscribing to related services. They could also learn and test knowledge on the protection against the COVID-19 via interactive Q&A. In

[37] https://www.who.int/health-topics/digital-health/dh-tag-membership.

[38] Rakuten Viber is a free messaging and calling app.

April 2020, WHO started Facebook-edition WHO Health Alert, which further enlarged the access to public health service.[39]

(2) Building an information cooperation platform and promoting global partnership in the fight against the COVID-19. On March 16, 2020, WHO launched the COVID-19 Partners Platform,[40] which served as a tool for all countries, program partners, donors and collaborators to collaborate in global COVID-19 response. The partner platform owned the function of real-time tracking and supported various countries in their precaution and response to the COVID-19. On May 5, 2020, WHO opened COVID-19 Supply Portal to collect supply requests from all countries and relevant partners and facilitate collaboration. Now, more than 125 countries and 50 donors have used the platform and portal.[41]

(3) Developing apps and promoting the consultation and sharing of knowledge on public health. On May 13, 2020, WHO launched an app called WHO Academy to support healthcare workers in the world. The app provided healthcare workers with relevant knowledge & resources that can be accessed on mobile devices, like the latest diagnostic guideline. In addition, WHO used the existing OpenWHO platform (openwho.org) to transform guiding opinions into training courses. Presently, the platform owns more than 2.5 million registered users and offers free training on 10 different topics in 22 languages. WHO also launched an app called WHO Info[42] to facilitate public concern with the latest action of WHO, the latest information on medicine and vaccines, and real-time data on the COVID-19.

### *5.4.2 Several European Countries Launched "Hackathon" Competition to Fight Against COVID-19 in a Coordinated Way*

In Europe, several governments develop innovative apps that meet various demands under the COVID-19 by building open innovation platforms. For example, netizens and enterprises co-launched "Hackathon" Competition themed on fighting against the COVID-19 in a coordinated way, via online channels. Now, eight countries have organized similar campaigns, i.e. Hack the Crisis in Estonia,[43] Wirvsvirus

---

[39] https://www.who.int/news-room/feature-stories/detail/who-launches-a-chatbot-powered-facebook-messenger-to-combat-covid-19-misinformation.

[40] https://covid-19-response.org/.

[41] https://covid-19-response.org/.

[42] https://www.who.int/news-room/detail/13-05-2020-launch-of-the-who-academy-and-the-who-info-mobile-applications.

[43] https://garage48.org/hackthecrisis.

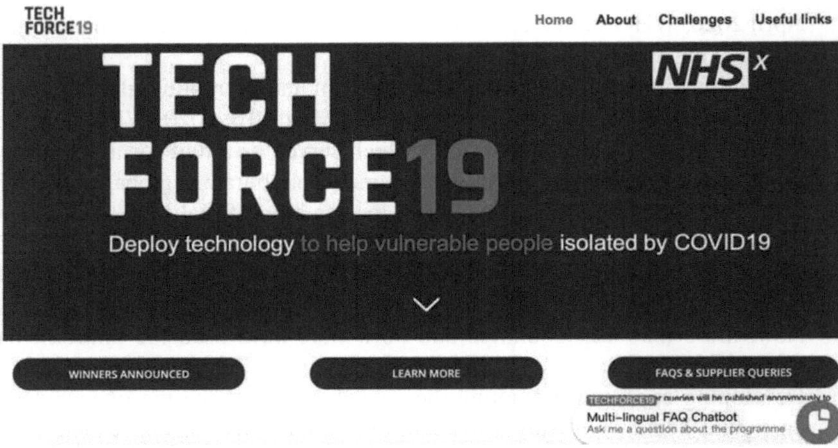


**Fig. 5.9** TechForce19 competition platform in the United Kingdom[52]

Hackathon in Germany,[44] Hack the Crisis in Finland,[45] Hack the Crisis in Lithuania,[46] Tech4Covid19 in Portugal,[47] Hack Force in Latvia,[48] Hack the Crisis in Poland[49] and TechForce19 in the United Kingdom.[50,51]

For example, in Estonia, Hack the Crisis was widely echoed shortly after its start, with more than 1,000 participants across 14 time zones in 48 h. In the United Kingdom, TechForce19 Competition (see Fig. 5.9) concentrated on the public in need of mental-health support and social care. Topics were providing telecare, volunteer service, etc. After it started, more than 1,600 people applied in 10 days.

Meanwhile, large-scale technological corporations, including Facebook, Microsoft and WeChat, worked with World Health Organization and launched "COVID-19 Global Hackathon" Competition. They also developed software to manage information on the COVID-19, covering such topics as health, business, education and entertainment.[53]

---

44 https://wirvsvirushackathon.org/.

45 https://www.hel.fi/uutiset/fi/kaupunginkanslia/pystytko-sina-pelastamaan-suomen-48-tunnissa-perjantaina-kaynnistyy-ennennakematon-haaste.

46 https://hackthecrisis.lt/.

47 https://tech4covid19.org/.

48 https://www.facebook.com/events/611058086114849/.

49 https://www.workinestonia.com/estonia-created-suve-a-state-approved-automated-chatbot-to-provide-trustworthy-information-during-the-covid-19-situation/.

50 https://techforce19.uk/.

51 https://www.nhsx.nhs.uk/covid-19-response/social-care/.

52 https://techforce19.uk/.

53 https://xw.qq.com/cmsid/20200423A0DO8R00.

### 5.4.3 *Germany and Italy Evened Out Personal Travel Tracking and Privacy Protection*

As EU's *General Data Protection Regulation* (GDPR) requires, EU countries commonly emphasize the protection of personal privacy in information acquisition and sharing in the prevention and control of the COVID-19. In order to even out the need for data and the protection of personal privacy in the fight against the COVID-19, EU member states took the measurements as follows, in the design of anti-COVID-19 products.

(1) Finding and notifying close contacts in an anonymous way. Robert Koch Institute (RKI) from Germany played a leading role in launching "Pan-European Privacy-Preserving Proximity Tracing" (PEPP-PT) Project,[54,55] and developing a mobile app that tracked close contacts via bluetooth detection technology. Daily contact information between users and others would be stored in "local files" of mobile phone after encryption. The stored information mainly contained the time and distance when users came into contact with others, rather than the information on users' real identity. Once one user was diagnosed as a confirmed case of the COVID-19, the mobile app would alert his/her close contacts.

(2) Raising the granularity of data sharing. By sharing collected mobile data of users with health departments, European mobile communication operators could oversee citizens' observance of quarantine requirements. For example, Telecom Italia, Vodafone and WindTre provided local governments with collected mobile data and helped to map major areas with dense COVID-19 and overall trajectory of users.[56] Collected big-granularity data provided governments with timely and accurate information and assisted them in making decisions. Simultaneously, the data did not involve individual information, which avoided the disclosure of data on personal privacy. In France and Spain, the data on users' health collected for the prevention & control of the COVID-19 were restricted to the use in public health departments.[57]

(3) Emergency legislation on privacy protection. Several European countries detailed requirements on information sharing and privacy protection related to the COVID-19. On March 9, 2020, Italian government issued a decree that authorized data sharing between public health department and civil defense department to deal with emergencies.[58] Italy also passed emergency legislation, requiring that whoever stayed in an area with the risk of the COVID-19 must

---

[54] https://www.sohu.com/a/385157208_161795.

[55] https://www.pepp-pt.org/.

[56] https://www.reuters.com/article/us-health-coronavirus-europe-telecoms/european-mobile-operators-share-data-for-coronavirus-fight-idUSKBN2152C2.

[57] https://tisi.org/13613.

[58] Source: *To Even out Decisions on "Privacy" and "Public Health": 10 Consensus, Differences and Challenges on the Protection of Personal Information in Various Countries in the COVID-19*. See https://tisi.org/13613.

notify health department.[59] Germany revised specific wording in the authorization rules of *General Data Protection Regulation* (GDPR) and empowered government to deal with personal data in the outbreak of pandemic, natural disaster and man-made disaster. Finland allowed the use of personal data to combat the COVID-19, and would enact laws to legalize the use of personal data for the prevention and control of serious infectious diseases.[60] By March 2020, nearly all Data Protection Agencies (DPA) in EU member states had released statements on data privacy in the COVID-19.[61]

### *5.4.4 Singapore Carried Out Data-Based Meticulous Management in the Fight Against COVID-19*

At the early stage of the COVID-19, new confirmed cases were limited, with no death cases, even though Singapore did not require common citizens to wear masks, stop work, close schools and close the city. Therefore, Singapore was extolled a "model" for the fight against the COVID-19 by World Health Organization and others.[62] Admittedly, as international pressure from the COVID-19 increased dramatically, the number of new confirmed cases rose in Singapore after April 2020. Singapore's experience, like meticulous management and data-based scientific decision-making, remains significant in post-COVID-19 era.

(1) Adopting a meticulous social management model and accurately tracking the COVID-19 infected cases. After the first case of the COVID-19 was confirmed on January 22, 2020, Singapore thoroughly surveyed the identity, route of infection and trajectory of activity of each case. In March 2020, GovTech of Singapore Government worked with Ministry of Health and launched a mobile app called "TraceTogether". It recorded the user's close contacts in the past 21 days via mobile phone-bluetooth signal interaction. It could determine people who came into contact with the confirmed patient and monitor & quarantine those close contacts via a precise tracking system[63,64], in order to curb the spread of COVID-19.

(2) Integrating data with traditional e-government facility and preventing & controlling the COVID-19 in an all-round way. In 2014, GovTech of Singapore government launched "Ask Jamie", an intelligent robot for governmental

[59] http://m.techweb.com.cn/marticle/2020-03-11/2780871.shtml.

[60] https://tietosuoja.fi/artikkeli/-/asset_publisher/tietosuoja-ja-koronaviruksen-leviamisen-hillit seminen?_101_INSTANCE_ajcbJYZLUABn_languageId=en_US.

[61] http://m.cheaa.com/n_detail/w_570175.html.

[62] Source: *What Makes Singapore Succeed in the Fight against the COVID-19 Naturally*, Xinhua Daily Telegraph. See http://www.xinhuanet.com/mrdx/2020-03/14/c_138876153.htm.

[63] https://www.xinjiapo.news/news/22180.

[64] http://paper.people.com.cn/rmlt/html/2020-04/10/content_1981760.htm.

website, which was put into effect on 70 governmental websites. In the COVID-19, Singapore government diversified chatterbot's function to provide consultation related to the COVID-19, and analyzed the interactive data on chatterbot to learn the most popular topics. In addition to Gov.sg and the website of Ministry of Health, citizens could communicate with the chatterbot via Facebook and Telegram.

### *5.4.5 South Korea Positively Developed Mobile Apps in the Outbreak of COVID-19 and Provided Effective Services*

In mid-late February 2020, the COVID-19 in South Korea was grave. To combat the COVID-19, South Korea strengthened capabilities in detection, treatment and management, and provided citizens with timely and customized services, with mobile apps playing a remarkable role.

In tracking and quarantining the COVID-19 patients, South Korea developed two apps, "Self-Diagnosis" and "Self-Quarantine". "Self-Diagnosis" App was for those who came to South Korea from abroad. They must download the app on their mobile phones, fill in their personal contact information and submit personal health information daily, so that government could master their health condition.

"Self-Quarantine" App was mainly used to monitor the location of quarantined people. On condition that the location of quarantined people went beyond the quarantine requirement or quarantined people showed relevant symptoms, both quarantined people and the staff of administration & security department in charge of supervision would be alerted simultaneously.[65] In this way, remote control could be achieved over quarantined people. On March 7, 2020, "Self-Quarantine" App was firstly used in Gyeongsangbuk-do and Daegu, where the COVID-19 was more severe.

South Korean government also organized or developed various mobile apps in remote education, pharmacy positioning and employment & office. "Now and Here" App calculated for users the risk in the area near the commuting route and decided whether users' route and confirmed patients' route intersected. If the user and patient appeared at the same place simultaneously, "Now and Here" App would alert the user and give him/her advice. Besides, South Korean government developed a special R&D app, which enabled citizens to retrieve the location of pharmacies and the inventory of masks in a real-time way (see Fig. 5.10).

[65] https://finance.sina.cn/china/gjcj/2020-04-19/detail-iirczymi7127173.d.html?from=wap.

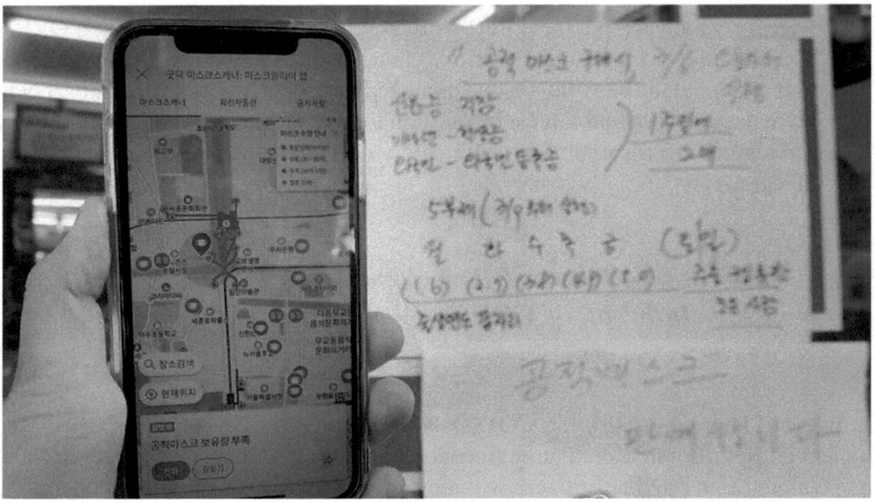

**Fig. 5.10** App used to retrieve the location of pharmacies and the inventory of masks in a real-time way[66]

# References

González, L., & Ruggia, R. (2020). Controlling compliance of collaborative business processes through an integration platform within an E-Government scenario. In *Proceedings of the 53rd Hawaii International Conference on System Sciences.*

González, L., Echevarría, A., Morales, D., & Ruggia, R. (2016). An E-Government interoperability platform supporting personal data protection regulations. *CLEI Electronic Journal, 19*(2), 8–8.

Helander, N., Jussila, J., Bal, A., Sillanpää, V., Paunu, A., Ammirato, S., et al. (2020). Co-creating digital government service: An activity theory perspective. In *Proceedings of the 53rd Hawaii International Conference on System Sciences.*

Koster, F., & Borgman, H. (2020). *New kid on the block! Understanding Blockchain adoption in the public sector*. https://doi.org/10.24251/HICSS.2020.219.

Liu, L., & Yang, D. (2020). Research on the reform of Blockchain-embedded governmental management model. *Administration Reform, 04*, 37–46.

Sánchez, L., Lanza, J., & Muñoz, L. (2020). *From the Internet of Things to the social innovation and the economy of data* (pp. 1–15). Wireless Personal Communications.

Tan, F., Hopkins, E., Chan, C., Leong, C., & Wright, A. (2020). *Digital platform-enabled community development: A case study of a private-public partnership sustainability initiative.* https://doi.org/10.24251/HICSS.2020.282.

Usman, M., Jan, M. A., He, X., & Chen, J. (2019). A survey on big multimedia data processing and management in smart cities. *ACM Computing Surveys (CSUR), 52*(3), 1–29.

[66] http://www.rmhb.com.cn/zt/zt2020/20200116_fyyq/202003/t20200320_800197905.html.

# Chapter 6
# The Development of World Internet Media

## 6.1 Outline

In 2020, a series of significant changes took place in the field of global Internet media. The evolution of communication technology and the change of international patterns jointly shaped the developmental pattern of global Internet media. In terms of audience, the audience transfer and content model of traditional social media changed remarkably. This aroused new-round product innovation among global digital giants. In terms of content, cross-border integration in streaming media platforms deepened, which caused a new-round competition among digital platforms. In terms of technology, the development of digital-media technology extended global Internet cooperation and produced computational propaganda represented by social-media robot. This stimulated the change of global Internet ecology.

In addition, the biggest challenge to Internet today is the sudden outbreak of the COVID-19 globally. It not only threatens the security of human society, but also impacts profoundly on Internet space. Scientists and medical personnel in various countries use Internet and cooperate in cyberspace globally, providing more possibilities for the control over the COVID-19. Clash and cooperation between countries co-exist in Internet space. "Infodemic" triggers numerous fake news in social media and the spread of malicious intention, and influences the trend of international relations to some degree.

As the global Internet pattern evolves, the definition of Internet media constantly changes with technological development. New technologies and new business forms alter and enlarge the connotation and extension of Internet media. Technologies such as wearable devices, sensor media, smart home, IoT and cloud storage continually develop and largely expand the boundary of media as an idea. Media-human relationship deepens. And the idea that "all serve as media when human and machine integrate" gradually comes true.



Chinese Academy of Cyberspace Studies, *World Internet Development Report 2020*,
https://doi.org/10.1007/978-981-16-9388-5_6

## 6.2 The Developmental Pattern of World Internet Media

In 2020, world Internet media presented a trend of diversified development. In the field of content production, new digital media platforms continued to emerge, which boosted the advance of content production in the context of omni-media. In the field of media consumption, the market share of digital publishing and online literature kept growing, and digital entertainment became a hot topic in online media consumption in the COVID-19. In the field of communicative technology, new-generation information technologies represented by 5G, AI and IoT further extended their use cases and tried to establish business models. These all greatly impacted on the developmental pattern of global Internet media in the future.

### *6.2.1 Digital Media Presents Diverse Developmental Trends*

1. Mainstream Digital Media

In 2019, the number of global mobile-end users exceeded 5.1 billion, accounting for 67% of the total world population. The number of Internet users exceeded 4.3 billion, accounting for 57% of the total world population. In particular, the number of active social media users reached 3.4 billion (see Fig. 6.1). North America maintained the highest penetration rate in network (95%) and social media (70%) in the world.[1]

Several traditional Internet giants remain the dominant role in digital media. Facebook, YouTube, and WhatsApp ranked top 3 in the number of daily active users in global social media in 2019 (see Fig. 6.2). Chinese Internet platforms played an equally important role in global social media, with the vast-market advantage and the ever-growing overseas strategy. WeChat, QQ and Sina Weibo ranked the 5th, 7th and 10th in the number of daily active users in global social media. Traditional social media like Twitter and LinkedIn ranked the 12th and the 14th in the number of daily active users respectively, with a slow trend of user growth.

Structurally, in 2019, global social media pattern became more vertical and segmented. In the summer of 2018, TikTok entered overseas market and became the second largest App in Google Play in only more than 12 months. It realized a download of 738 million times, only after WhatsApp (894 million).[2] Pinterest also developed rapidly and became one of the world's largest photo-sharing social websites. Noticeably, to tap segmented fields deeply and improve users' experience will play a decisive role in "diversified competition" among social media platforms in the future.

---

[1] Source: *ICANN Report 2019*. See https://www.icann.org/en/system/files/files/annual-report-2019-en.pdf.

[2] Source: TikTok Annual Download Report.
See https://www.statista.com/statistics/1089420/tiktok-annual-first-time- installs/.

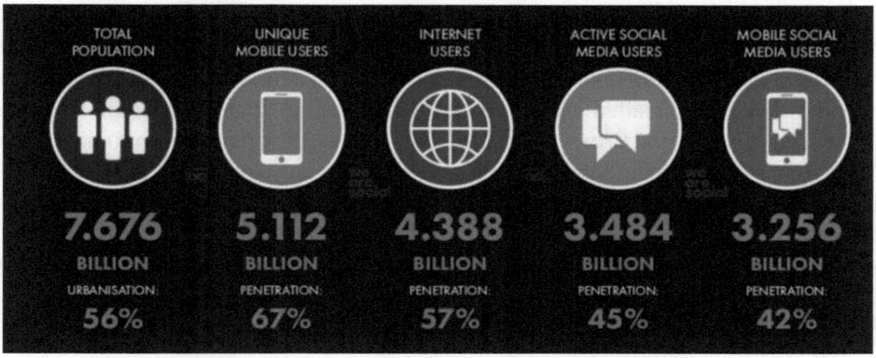


**Fig. 6.1** Global internet and social media user scale in 2019 (Unit: 1 Billion)

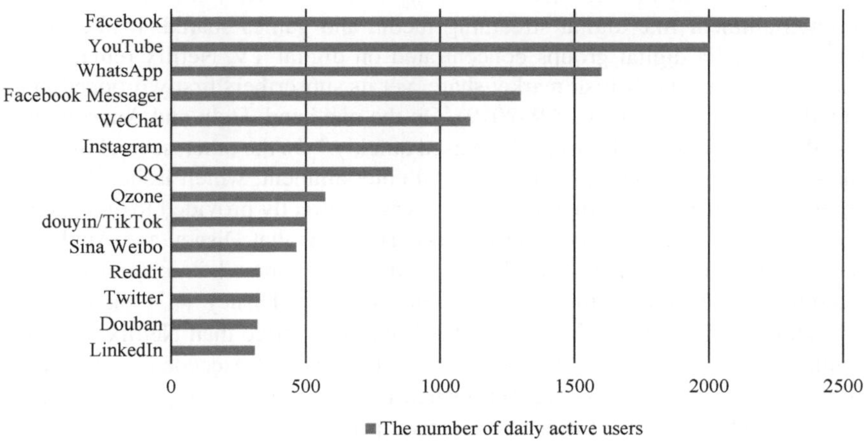


**Fig. 6.2** Ranking of global daily active users on social media in 2019 (Unit: 1 Billion)

In recent years, Instagram, a popular social media platform among teenagers, firstly grew by less than 10% in the number of American users, dropping from 10.1% in the end of 2018 to 6.7% in the end of 2019. The number of young users (25–34 years old) with strong consumptive power grew more than expected. Yet, it was hard for Instagram to maintain high-speed growth in user scale, under the pressure of the recovery of Snapchat and the rise of TikTok, two similar products. To cope with that, Instagram updated its Instagram Explore and delivered content to users more precisely via algorithms, which enabled advertisers to reach their target audiences more accurately. After the adjustment, Instagram realized a high-speed growth (over 10%) in advertising revenue in 2019.[3]

[3] Source: Emarketer Survey. See https://www.emarketer.com/content/instagram-user-growth-in-the-us-will-drop-to-single-digits-for-the-first-time.

In news app, Facebook, YouTube and Twitter remain three most influential social media news platforms in the world. Yet, emerging platforms in news industry like LinkedIn, Reddit and Snapchat performed averagely. Meanwhile, users' trust in the news of social media tailed off remarkably. In particular, after Presidential Election in the United States and Brexit in the United Kingdom, the fake news on social media resulted in sharp downturn in users' trust. As Pew's survey suggested, 59% of American adults did not think Facebook news was trustworthy and 48% of respondents expressed distrust of Twitter. 62% of respondents worried that excessive power of social media would cause many social problems, which meant a severe challenge to the reputation and sustainability of social media.[4]

2. Streaming Media and Digital Entertainment

Affected by the COVID-19, digital entertainment thrived in 2020. According to the research data of *The New York Times*, Netflix, a giant in streaming media, grew by 16% in flow in the COVID-19.[5] Owing to home quarantine, public demand for entertainment like digital streaming media and games soared. In fact, major streaming media digital groups concentrated on digital TV. Netflix remained the first in American and British market share, yet its subscribers firstly dropped in the United States in ten years (2009–2019). On the one hand, Hulu and Amazon challenged Netflix, whose subscribers increased quickly.[6] On the other, several industrial giants deployed or would deploy in digital entertainment, which lashed Over The Top (OTT) industry (i.e., streaming media service directly provided to audiences in cable or digital ways by service providers). The steps that Disney and Apple took were notable. In November 2019, Disney launched "Disney + ", a "family Union" streaming media platform that received wide attention. "Disney + " took advantage of Disney's multiple huge-flow IPs and offered lower price than Netflix. Only one month after its start, the total subscriber of "Disney + " exceeded more than 10 million.[7] In November 2019, Apple opened its streaming media platform, "Apple TV + ", and tried to attract users with a lower price.

In 2019, global game industry maintained high growth. Mobile phone games, online games and downloadable games ranked top 3 in market share. Specifically, the number of mobile phone game users reached 1.36 billion, and 25–36-aged users accounted for 36%. Predictably, games based on mobile-phone end would become the most competitive force in the entire game market in the next years.[8]

---

[4] Source: Statistical Data of Pew Research Center. See https://www.journalism.org/2020/01/29/an-oasis-of-bipartisanship-republicans-and-democrats-distrust-social-media-sites-for-political-and-election-news/.

[5] Source: Report of *The New York Times*. See https://www.nytimes.com/interactive/2020/04/07/technology/coronavirus-internet-use.html.

[6] Source: New Report at Forbes website. See https://www.forbes.com/sites/danafeldman/2019/08/21/netflix-is-expected-to-lose-us-share-as-rivals-gain/#7b32168d66d6.

[7] Source: Report of *The Verge*, a media-convergence news website. See https://www.theverge.com/2019/8/6/20757626/disney-plus-espn-hulu-bundle-price-date-streaming-service.

[8] Source: GoldenCasinoNews, Mobile Gaming Generated 60% of the Global Video Games Revenue in 2019,

In global mobile phone game market, duo-competition between China and the United States began to take shape. After 2018, China became the largest mobile phone game market in the world. In 2019, China realized a total revenue of $18 billion in the mobile phone game market and remained the first in the world. In contrast to the booming Chinese game market, the entire American market grew slowly, up by 2% year on year. In 2019, the United States realized a total revenue of $9.9 billion in mobile phone games.

The digital transformation of the global music market also became irresistible. In 2019, on-demand digital music hit a record high of 1.15 trillion times, up by 13.5% in streaming-media album sales over 2018 (by 32% in audio and by 10.6% in video).[9] This fully actualized the upgrading of the communicative channel in music industry. Similar to the music industry, digital platform-based Podcast developed quickly. Interactive Advertising Bureau (IAB) released *2019 Podcast Revenue Report*, which predicted that the advertising revenue of Podcast industry in the United States would exceed $1 billion in 2021.[10] In particular, in the COVID-19, Podcast industry grew fast, owing to citizens' long-time home quarantine or teleworking habit. Spotify, a music streaming media platform, monitored Podcast flow in the COVID-19, revealing that 19% of monthly active users listened to podcasts, up by 16% after the COVID-19. Podcast started to cooperate with audiobooks. Pushkin, an American startup podcast producer, introduced audiobook subscription into podcast and diversified the content for audiences.

3. Digital Publishing & Network Literature

In 2019, digital publishing industry was segmented significantly, with remarkable growth in digital audiobooks. In 2019, the United States realized an overall revenue of $983.3 million in the e-book market, down by 4.2% over 2018. Yet, digital audiobooks bucked the trend and grossed $576.9 million, up by 22.1% over 2018. In 2019, Simon and Schuster, a publishing giant, grew by 15% in global audiobook sales over 2018, in contrast to 1% in traditional e-books. That was true for Harper Collins, another publishing giant, whose published audiobooks increased by around 40% over 2018, with a strong rising trend.[11]

Dynamic digital publishers concentrated on content end and delivery, and more publishers tried to break the monopoly from digital giants (like Amazon's Kindle) on publishing platforms. Beat, a Norwegian technological corporation, co-forged Fabel,[12] an e-book subscription platform, with publishing giants Gyldendall and

---

https://goldencasinonews.com/blog/2019/12/30/mobile-gaming-generated-60-of-the-global-video-games-revenue-in-2019/

[9] Source: *The BuzzAngle Music Report 2019*, The BuzzAngle Music 2019 Year-End Report, https://rainnews.com/buzzangle-music-2019-report-the-u-s-generated-more-than-a-trillion-on-demand-streams/

[10] Source: Nieman Lab Report, Nicholas Quah, Can Pushkin Industries bring the podcast and audiobook audiences together?, https://www.niemanlab.org/2020/05/can-pushkin-industries-bring-the-podcast-and-audiobook-audiences-together/

[11] Source: Statistics from the Website of Goodreader Publishing, Kozlowski (2020).

[12] Source: *Publishing Perspectives* News, Anderson (2019).

Aschehoug. VK, Russia's largest social media platform, also worked with publishers to create a Russian-language e-book market.

In global digital publishing and content production, Chinese network literature displayed a unique style and formed the highlight in the global digital content market. In 2019, translated into more than 10 languages, Chinese network literature reached more than 40 countries and regions along "The Belt and Road". According to *2019 Chinese Network Literature Going Global Research Report* released by iResearch, 40% of readers at overseas online articles would like to pay. The Report predicted that the number of potential users of overseas Chinese network literature in Southeast Asia, Europe and the Americas would exceed 800 million, which would create an over-30 billion-*yuan* market with the main body of paid reading, advertising and copyright operation.

### 6.2.2 The Use Cases of Cutting-Edge Technologies Become Increasingly Diverse

1. 5G Diversifies the Information Form in Media Industry

In recent years, the maturity and popularization of 5G technology stimulate new media forms and change human information communicative ecology. The high-speed transmission capacity of 5G transforms and upgrades the form of content production. 5G is featured with super-large channel capacity and GB-based high-speed transmission, with its practical speed in user experience growing exponentially. The breaking of technological barriers further boosts the turn from a text & image-dominated platform to an audio & video-dominated platform in the mobile Internet. The rise of TikTok in the global market produces a "catfish effect" and propels traditional social media like Facebook and Twitter to propagate their short-video content platforms. High-speed, high-definition native video content production that is driven by 5G becomes a new competitive field for global social media giants.

5G technology is gradually actualized, which enables VR/AR industry to recover after market chill. The high speed and low delay of 5G not only meet the transmission demand of VR videos, but also improve the definition and fluency of videos greatly. This ensures high-quality immersive experience of VR/AR. As the report of Intel estimates, next-generation immersive VR/AR apps will come true by 2025, and AR games will account for more than 90% of "5G + AR" revenue by 2028.[13] The optimistic outlook on VR/AR stems from the wide deployment of 5G technology in Europe, the Americas and China recently. As 5G technology is gradually actualized, the case & emotion- oriented feature of VR/AR further emerges, with "immersive communication" as a common form in the 5G era.

[13] Source: *Forbes* News. See https://www.forbes.com/sites/solrogers/2019/01/30/the-arrival-of-5g-will-unlock-the-full-potential-of-vr-and-ar/#6ea5b7157bcc.

5G technology re-maps the form of the game market. "Cloud Gaming" probably becomes more popular as 5G technology develops. "Cloud Gaming" refers to a new form that games can be directly run at cloud-end servers with no restrictions on terminal. It is no configuration, no download and cross-terminal. Presently, Amazon, Sony and Microsoft start to deploy "Cloud Gaming". The game field with a wide market and mature business models possibly becomes an important use case after 5G is commercialized.

2. AI Promotes the "Intelligent Upgrading" of Media Industry in an All-Round Way

With technological innovation and capital entry, AI technology is gradually used in media industry, creating three use cases, i.e., sound-controlled intelligence media, intelligent advertising and automated fact checking.

"Sound-controlled intelligence media" represented by intelligent loudspeakers becomes an important commercial field in the development of AI. By the end of 2019, global intelligent loudspeaker sales reached 147 million units, up by 70% over 2018.[14] Technological and Internet corporations like Amazon, Google, Apple, JD, Xiaomi and Baidu successively launch their intelligent loudspeakers and strive to seize the opportunity in human–computer interaction in future smart home. As intelligent loudspeakers become more popular, the potential of Podcast will fully emerge in the future. Traditional broadcasting corporations like BBC (UK), NPR (US) and Australian Broadcasting Corporation (ABC) also start to invest in on-demand podcast news services to solidify their market shares and actively explore customized on-demand content for intelligent sound devices.

The intelligent media ecosystem affects advertising industry. Internet enterprises increasingly apply intelligent advertising into the delivery of advertising content, which can dynamically allocate various data and advertising material libraries and ultimately create advertising for users in intelligent advertising platforms, according to users' personalized information and media-use environment as well as algorithm. Since 2019, intelligent advertising has been widely used in digital platforms. Spotify delivers different types of advertising content to users in a customized way, in the light of their music-listening preferences. In advertising delivery, to accurately determine users' needs and obtain massive users' information by using big data and AI forms an important means for AI digital marketing in the future.

In journalism, AI technology is widely used in both content production and fact checking. Automatic fact-checking uses AI deep learning to automatically check factual elements in news and substantially screen false and misleading information on the Internet. In 2019, researchers from University of Waterloo (Canada) developed an automatic fact-checking tool, which used deep-learning algorithms to judge whether claims made in one piece of news were supported by other articles on similar topics or not. If the content of the news was not supported by mainstream media or articles

---

[14] Source: *Business Wire* Report. See https://www.businesswire.com/news/home/20200213005737/en/Strategy-Analytics-New-Record-Smart-Speakers-Global.

from authoritative sources, the news would be judged as suspicious information.[15] The technology gradually matures and helps social media to screen false information fabricated by users, which effectively purifies Internet information environment.

3. IoT Enhances the Integration of Media and Reality

Now, IoT develops explosively, and its popularity promotes the development of media industry remarkably. On the one hand, IoT can provide massive data, so that the owner of IoT-end can learn a user's trend quickly. Apple Watch (Apple) can detect the data on a user's current heart rate and record resting heart rate, walking heart rate and breathing heart rate in a real-time way. This not only helps users to monitor their physical state at all times, but also makes it possible to master overall health state of users in a region with macro data. The connectivity mechanism with man–machine symbiosis provides abundant data and information to IoT device-end producers and helps to make decisions on expanding industrial chains and innovating industrial layout. Owners of IoT and other related technologies will exert new impacts on the structure of media industry. On the other hand, the use of IoT in fields like automobile driving, family life and personal health will promote the wider distribution of communicative terminal, with intelligent devices as a medium. Automobile corporations represented by Tesla benefit from the transformative impact of IoT technology and establish the IoV system on roads. As Tesla conceives, after IoV becomes popular, every intelligent car serves as a "medium", which forms a data network of car and road. With the large IoV system, real-time interaction can be achieved, and all-round coordination can be done among intelligent traffic management, intelligent dynamic information service and intelligent vehicle control. As above cases indicate, the development from Internet to IoT accelerates the digital use in everyone's life, or impacts greatly on human social forms.

## 6.3 The Clash and Cooperation Among World Internet Media Co-Exist …

In the global fight against the COVID-19, Internet media plays an important role. "Infodemic" that results from Internet media exerts equally social influence with the COVID-19. The competition, clash and cooperation among countries in Internet media embody the complex structure of cyberspace. As a major event in human history, the COVID-19 helps us to fully understand the present pattern and future trend of global Internet media.

[15] Mok (2020).

### 6.3.1 "Infodemic" Comes to Rise in the Outbreak of COVID-19

In the global fight against the COVID-19, "infodemic" arouses wide attention among international organizations and various countries, like World Health Organization (WHO). The COVID-19 triggers the first real "infodemic" in social media, which disseminates misleading information in the world at an unprecedented speed. It forms a big challenge to various countries to cope with "infodemic" that spreads rumors and lies on the Internet and social media.

In home quarantine, citizens in various countries presented high-degree "media dependency". According to the statistics of APP Annie, from January to March in 2020, the time that global netizens spent on mobile phones increased by 20% year on year.[16] In a survey of Internet users in the United States and the United Kingdom, 68% of netizens usually searched and browsed news related to the COVID-19, which were the most frequent use of Internet in quarantine.[17]

Accordingly, BBC developed COVID-19 Dashboard, an "Epidemic Information Collection Map System", with Gisanddata, a geographic information & data platform. Gisanddata deeply collaborated with American universities. With the technological framework provided by two Chinese doctoral students and their supervisor at the Center for Systems Science and Engineering of Johns Hopkins University, Gisanddata used a multi-layer map data system to provide real-time visualization of the COVID-19 in the world and toolkit for real-time analysis of the COVID-19 via AI robot. This gave data to the public in the world, who wanted to know the development of the COVID-19.

By the end of June 2020, there were more than 600 million posts with hashtag "# COVID-19" or "# coronavirus" on Twitter, with more than 100 billion times of interaction on the topic. This marked the widest attention on the Internet since the birth of social media. Nevertheless, false information also came and formed "infodemic" that combined rumor, misleading information and fake news and spread on social media at an unprecedented speed.

As World Health Organization (WHO) defines in its report, "Too Much Information" and "False-True Mix" are prominent features of "infodemic" today. As the survey of Pew Research Center reveals, 71% of Americans receive too much information on the Internet in the COVID-19, and 43% of Americans state reading COVID-19-related information makes them feel blue. Nearly 50% of Americans say they cannot clearly judge true and false information in the COVID-19, and as much as 64% of Americans admit they encounter false information on the COVID-19.[18]

---

[16] Source: Statistical Report of App Annie, Venkatraman (2020).

[17] Source: Statistical Data of Visual capitalist website, Jones (2020).

[18] Source: Statistical Report of Pew Research Center, AMY MITCHELL, J. BAXTER OLIPHANT, ELISA SHEARER. About Seven-in-Ten U.S. Adults Say They Need to Take Breaks From COVID-19 News [https://www.journalism.org/2020/04/29/about-seven-in-ten-u-s-adults-say-they-need-to-take-breaks-from-covid-19-news/2020-4-29.

Many extremist views also spread quickly on the Internet, with the help of conspiracy on the origin of the virus. According to a survey done by BBC and Anti-Extremism Think Tank, many Far Right groups vilified immigrant groups and Muslim groups on Facebook, claiming that the COVID-19 virus was deliberately created and disseminated by the elites. Besides, the survey specially studied the Facebook homepages of 34 Far Right parties, revealing that they commonly posted or forwarded false information on the COVID-19, e.g., the claim that the global outbreak of the COVID-19 resulted from several-year human conspiracy. Meanwhile, the time of interaction on these pages was extremely amazing. For example, in the corresponding period, the time of interaction on the homepage of World Health Organization reached 6.2 million, and the total time of interaction on the homepages of 34 Far Right parties reached 80 million.[19]

In addition, a non-profit public information platform jointly established by media and academic institutions in the COVID-19 became the model case for the governance of infodemic. Social media platforms like Google, Facebook, and Twitter successively allocated large-amount budgets to promote information governance campaign, with authorities like World Health Organization and the United States Centers for Disease Control and Prevention (CDC). Both Facebook and Twitter introduced information screening & marking mechanism in the prominent places of their platforms. The post with "blue tick" meant the information was real and reliable, while the post with "small red flag" meant it contained misleading information. This helped users to recognize and judge. Major platforms strengthened the tracking and verification of "virus-like memes" that were forwarded in massive times. Any memes that were labelled as "misleading information" could be only browsed and not be shared.

Social platforms like Reddit and WhatsApp also enhanced the monitoring of chat groups. Groups that disseminated extremist views were "isolated" and their updating was ceased. Groups that spread massive misleading information or conspiracy were labelled with "small red flag" or with warnings "CONTAINING UNCONFIRMED, FALSE AND DANGER INFORMATION". In China, Tencent's Fact Check Platform specially opened a window called "COVID-19 Real-Time Counter-Rumor". It marked information that spread on social media as "True" "Unconfirmed" and "Rumor" and explained how the information was verified under each piece of news, to help users identify the authenticity of information on social media.

### *6.3.2 COVID-19 Gives Impetus to Global Digital Cooperation Course*

In the COVID-19, global cooperation promoted by social media and digital platforms became a new highlight that united the forces of all countries in the global fight against

[19] Source: BBC News Report, Miller (2020a, b).

the COVID-19. In medical cooperation, global coupling promoted by digital platforms boosted the flow and sharing of epidemic information and medical resources, which effectively helped the fight. Zoom, a video conference service provider, collaborated with more than 1,000 Chinese hospitals and achieved tele-consultation, patient diagnosis and treatment via high-quality video conferences.

To help enterprises turn the corner, social media and digital platforms also extended service scopes and improved & innovated their services globally. Major video conference service providers offered free service to global enterprises. In early March 2020, Microsoft Teams, an office app from Microsoft, was freely opened to all users, whose daily active users reached *c.* 75 million in early May. In early April 2020, DingTalk, an office app from China's Alibaba, launched its overseas edition DingTalk Lite, which was freely opened to global users and helped enterprises' teleworking. The COVID-19 greatly promoted the development of teleworking. Internet teleworking and online meeting probably become a new trend in the future.

In education, digital platforms like video conferencing provided free service to many schools to help mitigate COVID-19's impact on teaching. Microsoft worked with UNESCO and University of Cambridge and started Learning Passport, an online tele-learning app for children in remote, rural and refugee areas.[20] Google announced to provide G Suite users with free advanced group chat service to help teachers better organize online classes. On March 1, 2020, DingTalk released the Japanese-edition teleworking guide to help Japanese enterprises and schools work online.

In addition, social media and digital platforms helped international organizations to fulfill their functions. World Health Organization cooperated with Google and launched Google SOS Alert, which ensured users firstly saw information published by authorities like World Health Organization and the United States Centers for Disease Control and Prevention (CDC) when they searched keywords like "COVID-19", by optimizing algorithms. World Health Organization (WHO) signed a cooperation agreement with WhatsApp and delivered special information on the COVID-19, whose service covered the latest news on the COVID-19, travel advice and rumor judging. To ensure that global users could easily and quickly access authoritative information, these pages or links were all showed in English, Chinese, Arabic, Russian and other languages. Tencent Meeting, Tencent's online conference platform, reached an agreement with the United Nations, which supported the organization of more than 1,000 global online conferences.

Social media and digital platforms extended their fight against the COVID-19 to global netizens' cultural & entertainment life. Large-scale online entertainment activities relaxed citizens who in turn fought against the COVID-19 better. Simultaneously, charity performance could raise funds and help the global fight against the

[20] Source: Microsoft News Center. UNICEF and Microsoft launch global learning platform to help address COVID-19 education crisis. https://news.microsoft.com/2020/04/19/unicef-and-microsoft-launch-global-learning-platform-to-help-address-covid-19-education-crisis/

COVID-19. "One World: Together At Home", co-organized by World Health Organization and Global Citizenship (a NGO), was broadcast around 20 million times, raising nearly $128 million for healthcare personnel.[21]

### 6.3.3 Divergences on the Values of World Internet Media in the Outbreak of COVID-19

In face of COVID-19, a common challenge to human, world Internet media and institutions positively cooperated. Yet, global political, economic and cultural disharmony that arose from COVID-19 must be noticed. Various countries often displayed their stances when they faced the dual identity of "nation-state" and "citizen of the world".

In COVID-19, Western countries implemented protectionism and launched a series of confrontations in cyberspace, declaring the legitimacy of their governance strategies. This ignited large-scale debates. With the development of network technology and Internet media, issues like cyber sovereignty and information security came to rise. The legitimacy of global governance in cyberspace and the tendency of global public opinion impacted enormously on global Internet ecology.

In March 2020, two senators of US Congress wrote to the CEO of Twitter, "strongly urging" the latter to remove China's official account from Twitter.[22] In the letter, the two senators alleged that China was using Twitter to spread propaganda in a dangerous global crisis and demanded that the United States must ban Chinese government & officials from accessing US-owned social media platforms. From August to September in 2019, Twitter massively deleted more than 4,000 accounts related to China's Hong Kong issues in China. The ongoing actions of social media like Facebook and Twitter & American senators told that ideological and political-value boundary based on Internet media and social media started to change in COVID-19.

For a long time, the United States and Russia accused each other of using cyberspace to threaten their national security. The United States charged Russia with using social media to launch "cyberspace war" in 2016 US Presidential Election, which influenced the final election result. Russia's Internet Research Agency (IRA) had been regarded by the United States as Russia's main agency that executed cyberspace war and Internet attack. On the eve of 2020 Election, US Democratic Party officials specifically spoke of the need to watch out Russia's using Internet to interfere in US Election again.[23] In COVID-19, Indian and US governments demanded that Bytedance sold its TikTok business in India and the United States with the pretext of

---

[21] Source: NGO Global Citizenship Report, 'One World: Together At Home' Raised Almost $128 Million in Response to the COVID-19 Crisis. https://www.globalcitizen.org/en/content/one-world-together-at-home-impact/

[22] Source: American Political News Website, Miller (2020a, b).

[23] Source: Report of *The New York Times*, U.S. Escalates Online Attacks on Russia's Power Grid, https://www.nytimes. com/2019/06/15/us/politics/trump-cyber-russia-grid.html.

"endangering national security". India even banned some users' access to TikTok. Noticeably, despite that win–win cooperation in the global Internet field formed an irresistible trend, the silent rise of Internet protectionism could not be overlooked in COVID-19.

## 6.4 Hot Topics Among World Internet Media

In 2020, there were a series of hot topics that centered on Internet development. Computational propaganda represented by social media robots expanded in Internet space and impacted greatly on world Internet ecology. The trend of media platformization further developed, and the application of digital currency and big data deepened the impact of digital platforms on society. In global Internet cooperation, BRICS countries and countries along "The Belt and Road" collaborated in Internet in a closer way. The development of mobile social media resulted in the idea of "news hour" that was decided by the use case of media. To comb and analyze these hot topics reveals the new trend in world Internet, in terms of content, platform, ecology and international cooperation in the future.

### *6.4.1 Computational Propaganda Affects the Ecosystem of World Internet (Media)*

Computational propaganda refers to the use of Internet, particularly social media platforms, to collect, analyze and evaluate the targeted data on audiences. On the basis of this, software programs like intelligent robots serve as a propagandistic means that simulates human in information dissemination and online interaction to affect and shape public opinion. The main task of computational propaganda is to manipulate public opinion. It is one of the most recent and popular technological strategies adopted by those who want to use information technology for social control. As research suggests, computational propaganda has been widely used in politics and business in many Western countries. False account is a means that "paid Internet trolls" often use in computational propaganda. According to statistics, more than 80% of computational propaganda use real accounts ("paid Internet trolls") or robot accounts. And the content to be disseminated are mainly positive or negative information of specific political parties and politicians, the harassment message to specific users and personal attacks.[24]

Computational propaganda has threatened the fairness and effectiveness of political elections in various countries. The team led by Oxford Internet Institute discovered that in elections and political crises in the United States, Russia and Ukraine, there were signs of computational propaganda in social media with social media

[24] Bradshaw and Howard (2019).

robots. This suggested that computational propaganda extensively influenced political campaigns in various countries. We must attach more attention to how computational propaganda manoeuvres the dissemination on world Internet, for further observation and analysis.

More significantly, the influence of computational propaganda extends to fields like international communication and public diplomacy. As research shows, in top 100 most-tweeted accounts on the Twitter platform with China-related hashtags, none is "pro-China" in stance, yet more than 50% are "anti-China".[25] Similarly, in December 2019, China Global Television Network (CGTN) uploaded *Fighting Terrorism in Xinjiang* on its YouTube account with 1.01 million subscribers. However, the time of viewing was inconsistent with the number of subscribers, suggesting an intentional streaming-restriction in YouTube.

### 6.4.2 Digital Omni-Media Displays a Prominent Trend

In the Internet era, the development of media convergence presents a trend of integration, and omni-media comes to rise. Omni-media is viewed as a future form of digital media. Social media, including Google, Tencent, Amazon and Facebook, firstly ran based on specific objectives and business (shopping, social network, Internet search, etc.). Yet, as the scale and business of these platforms enlarged, social media became important infrastructure in world Internet and played a key connecting role in the flow of information and service. Their impact on real society grows increasingly, and some omni-media forms the power beyond nation-state.

It is viewed that Facebook's influence is larger than many countries. It rivals many governments, no matter in its ability in information, service and social organization or economic scale, population and involvement in public governance.[26] In June 2019, Facebook officially released *White Paper on Digital Cryptocurrency Project Libra (Diem)*. Facebook concentrates on digital currency, making progress in realizing "virtual" quasi-nation community. Libra is an important link of Facebook that promotes the development of platform. Having turned from social media to omni-media, Facebook hopes to enhance its impact on the real world with financial tools like Libra. Evidently, the fact that omni-media gradually penetrates into real life will foster the integration of various applications, functions, services and information in media platforms. Cooperating with governments, transnational corporations and financial institutions, omni-media builds platform media that covers the entire society.

In China, WeChat and Alipay develop from social media & payment tools to omni-media (as a trend) to large extent. Functionally, their social services reach the level of public governance. In particular, in COVID-19, WeChat and Alipay successively cooperated with General Office of the State Council and the State Information Center and launched their own "Health Code". Users could obtain their "Health Code" after

[25] Bolsover & Howard (2018).

[26] Xu and Yao (2018).

they logged in, which realized overall monitoring and real-time tracking of COVID-19 information. WeChat and Alipay come to daily society more deeply, which plays a significant role in the digitization of social structure and the networking of public governance.

### 6.4.3 Transnational Internet (Media) Cooperation Continually Deepens

Now, the Internet becomes important infrastructure for international community. Under such circumstance, Internet-based cooperation among countries keeps deepening. In October 2019, Brazil and China firstly proposed that BRICS countries jointly build a streaming media video platform, to create more developmental space for film and streaming media video industry and to break global monopoly of streaming media platforms like Netflix. This was the first initiative to establish the mechanism that ensured the diversity of global digital culture.[27]

Under the cooperative framework of "The Belt and Road" Initiative, cooperation among countries on the Internet and digital media continually deepens. Belt and Road Economic Information Partnership (BREIP) was founded in June 2019, by Xinhua News Agency and other 32 agencies, including Polska Agencja Prasowa (PAP), Class Editori (Italy), INTERFAX (Russia) and Azerbaijan State News Agency. BREIP is used for information exchange among member agencies. Member agencies can share real-time, free and dynamic information on investment, trade, industry, project and enterprise related to "The Belt and Road", which promotes deep cooperation on "The Belt and Road" digitally. Besides, Internet cooperation based on "The Belt and Road" framework, like "The Belt and Road" Information Port and Electronic World Trade Platform (eWTP) Hangzhou Pilot Area, keeps advancing.

"The Belt and Road" international cooperation based on next-generation communication technologies (like 5G) grows remarkably. Many European countries plan to sign cooperation agreements with 5G service providers like Huawei, and take the opportunity to join regional digital Internet cooperation along "The Belt and Road". This promotes global cooperation to a new stage, with the Internet and digital media technology as the core.

Simultaneously, North–South Cooperation between developed and developing countries advances. Digital Premium (Brazil) and Google (the United States) co-launched *Digital Premium Jornais*, which adopted "exhibition & marketing" cooperation in content. Google worked with 18 leading news & media enterprises in Brazil, established a technology-driven content exhibition & marketing platform, collected high-quality news & media resources and advertising resources across Brazil, and publicized them via Google's network platform (e.g., Google AdExchange). The giant-giant alliance built the platform into one of the most reliable sources of local

[27] Source: Tass News Report, Tass, BRICS challenges Netflix: Five-nation club seeks alternative to US film-streaming giant, https://tass.com/economy/1084916.

news in Brazil, whose content covered seven Brazilian states, with an average daily visit of 7 million and client-end users of more than 5 million. *Digital Premium Jornais* boasts a "national" platform for new & media resources in Brazil. The cooperative model successfully maximizes the utilization of news resources and boosts global re-delivery of news resources in the world Internet community.[28]

### 6.4.4 The Content Delivery of Internet Media Advances Toward Meticulousness

The wide popularity of the Internet and audiences' habits in digital media deeply alter the way news is produced and distributed. It means that information production also has its "biological clock", which requires us to produce and disseminate information by learning users' demand in different times in a day. Therefore, news delivery accords with the idea of "news hour".[29]

In *Reuters Institute Digital News Report 2019*, Research Institute for the Study of Journalism proposed that consumers of digital news represented by young audiences presented different demands for news in different cases in a day, summarized as "Four Key News Hours".

(1) Focus Hour. Often in evening or on weekend, when users are absorbed in information.
(2) Updating Hour. Often in morning, when users need to obtain key information quickly and efficiently.
(3) Leisure Hour. Users read information when they do something else.
(4) Break-off Hour. Users stop what they are doing to browse the delivered information.

As the consumption and use of news become more fragmented and closely involve the audience's cases, both traditional media and emerging news collection platforms start to include the idea of "news hour" in their news production and delivery. In 2019, BBC gradually introduced the idea of "news hour" in the news delivery on its social media. BBC newsroom delivered news content with various topics and types to users according to users' different times and cases, when it designed delivered content. This optimized the match between content and the audience's cases. If the content production (form) and delivery (time) matched the demands of the audience at specific hours, better communication effects would come out.

Therefore, Internet media must avoid "bombing" users with homogeneous content repeatedly. Rather, they need to consider cautiously how to integrate different types of information into a "story package" and deliver it precisely at different "news

[28] Shi and Dai (2020).

[29] Source: NiemanLab Report, Marshall, S. The year to learn about news moments. https://www.niemanlab.org/2020/01/the-year-to-learn-about-news-moments/

hours" and platforms. Presently, the delivery of Internet news impacts on the audience's way of reading news more deeply. As omni-information media continually develops, Internet media becomes more complete and meticulous. The delivery form represented by "news hour" better meets the psychological needs and use cases of the audience, which inspires us in the fast development of mobile social media today.

## References

Anderson. (2019). A publisher's own platform: Norway's beat technologies heads for frankfurt Audi. https://publishingperspectives.com/2019/09/at-frankfurt-audio-2019-frankfurter-buchmesse-norway-beat-technologies.

Bradshaw, S., & Howard P.N. (2019). The global disinformation order 2019 global inventory of organised social media manipulation. https://comprop.oii.ox.ac.uk/wp-content/uploads/sites/93/2019/09/CyberTroop-Report19.pdf.

Bolsover, G., & Howard, P. (2018). Chinese computational propaganda: automation, algorithms and the manipulation of information about Chinese politics on Twitter and Weibo. Information. *Communication & Society*, 1–18.

Jones, K. (2020). How COVID-19 has impacted media consumption, by generation [EB/OL]. Visual capitalist. https://www.visualcapitalist.com/media-consumption-covid-19/.

Kozlowski, M. (2020). Ebook revenue fell 4.2% in 2019 and generated $983.3 million. https://goodereader.com/blog/digital-publishing/ebook-revenue-fell-4-2-in-2019-and-generated-983-3-million.

Miller, C. (2020a). Coronavirus: far-right spreads Covid-19 'infodemic' on Facebook. BBC News https://www.bbc.com/news/technology-52490430.

Miller, M. (2020b). GOP lawmakers call on Twitter to ban Chinese Communist Party from the platform. The Hill. https://thehill.com/policy/cybersecurity/488731-republican-lawmakers-call-on-twitter-to-ban-chinese-communist-party-from2020b.3.20.

Mok (2020) Deep learning AI tool identifies 'fake news' with automated fact checking. https://thenewstack.io/deep-learning-ai-tool-identifies-fake-news-with-automated-fact-checking/.

Shi, A., & Dai, R. (2020). Revitalizing local news in the era of intelligent media: Path & model. *Youth Journalist*, 2020(04)

Venkatraman, A. (2020). Weekly time spent in apps grows 20% year over year as people hunker down at home. https://www.appannie.com/en/insights/market-data/weekly-time-spent-in-apps-grows-20-year-over-year-as-people-hunker-down-at-home/.

Xu, C., & Yao. J. (2018) Is facebook a country: 'Facebookistan' & nationalized imagination in social media. *Shanghai Journalism Review*, 2018(11):15–25.

# Chapter 7
# The Development of World Cybersecurity

## 7.1 Outline

Presently, new technology and new application like big data, cloud computing, AI, 5G and IoT breed new forms of digital economy. Digital form, platform-oriented form, intelligent form and ecological form re-forge society. Cyberspace has become a major field in which leading countries contend with each other. In 2020, the outbreak of COVID-19 enormously impacted on world cybersecurity, which signaled a new stage for world cybersecurity development.

In terms of world cybersecurity, the situation in cyberspace remains grim, and APT attack becomes one of the biggest threats to the strategic stability of cyberspace. Incidents on data leakage keep rising, the vulnerabilities of and threats to open source software aggravate the situation, and hackers take advantage of the vulnerabilities on cybersecurity to launch cyberattack. The data cloud of corporate business intensifies the risk of cloud-computing security. Under the impact of COVID-19, online business and medical service see severe cyberattack.

Various governments in the world promptly establish and improve strategic policies on cybersecurity, and continually strengthen their capabilities on cybersecurity, emphasizing data security and new-technology fields. Particularly, since the outbreak of COVID-19, many countries have remarkably enhanced the protection of personal privacy. The scale of global cybersecurity industry grows steadily, in which North America, Western Europe and the Asia–Pacific Region equally maintain a leading position. Besides, market demand keeps enlarging, and service market develops in a favorable way.

Many countries view the training of talents on cybersecurity as a strategic measure that improves national comprehensive capacity in cybersecurity. Even though fruitful achievements are made, the need for talents on cybersecurity is insufficiently met. This poses a great challenge to cybersecurity work now and in the future.



Chinese Academy of Cyberspace Studies, *World Internet Development Report 2020*,
https://doi.org/10.1007/978-981-16-9388-5_7

## 7.2 World Cybersecurity Develops into a New Stage

The evolution of cyberspace quickens and the competition among major countries intensifies. Under such circumstance, COVID-19 overspreads globally and causes many changes in cybersecurity. Generally speaking, world cybersecurity remained extremely severe in 2020.

### *7.2.1 The Evolution of Cyberspace Results in New Changes in Cybersecurity*

1. The Evolution of Cyberspace Technology Exacerbates the Threats to Cyberspace

As new technology and new business like AI, big data and blockchain develop rapidly, cyberspace displays a digital, networking and intelligent trend, which puts more pressure on the protection of cybersecurity. On the one hand, in new technology and new business, the technological system and business management do not mature and face the risk of vulnerabilities on cybersecurity. In recent years, new technological vulnerabilities of AI and blockchain have been substantially utilized. On the other hand, new technology and new business result in new difficulties for the protection of cybersecurity, which sharpens the excessive collection of data and the data leakage of personal privacy. In addition, the abuse of such technologies as deepfake and algorithmic recommendation probably endangers national security in a deep way. Meanwhile, the empowering effect of cyberspace poses more complex challenges to cybersecurity.

2. The denotative effect of cyberspace enlarges the connotation of cybersecurity

Currently, human face a turn from information society to intelligent society. Innovations are made in new technology and new business like broadband internet, IoT, cloud computing, AI and life science, which realizes deep integration among physical space, cyberspace and biological space. This radically impacts on human society. The object of cybersecurity protection extends from traditional computers and servers to various terminal devices like cloud platform and big data platform. The boundary of cybersecurity keeps expanding, and the scope of cybersecurity protection keeps enlarging. The widespread distribution or existence of cyber vulnerabilities raises the cost of cybersecurity protection and produces greater pressure. The universality of attack target and the incomprehensiveness of protection create lots of chances for attackers. Besides, the anonymity in cyberspace conceals the hidden danger and increases the difficulty of active defense.

### 7.2.2 The Game Among Major Powers (Competition & Cooperation) Poses New Challenges to Cybersecurity

1. Cybersecurity plays an increasingly important role in the competition or game among major powers

The confrontation in cyberspace means an extension of human-intelligence confrontation in virtual space. In cyberspace, the relations among major powers mirror and extend their game in physical space, in which the dispute on cybersecurity probably deteriorates into conflict in the real world. Cybersecurity functions as not only a technological idea but also the means that countries take to contain their strategic competitors in their confrontation. In cyberspace, some countries take preemptive measures and active steps, combine cyberwar with traditional military attack, and implement supply-chain outages and technological blockades, on the pretext of protecting cybersecurity. They even curb foreign corporations with national power, which interdicts normal free trade and transforms cybersecurity into a new barrier for global free trade. Several countries turn a blind eye to the international division of labor that takes shape under the background of globalization in recent years and vigorously carry out the domestic-targeted return of industrial chain (or the so-called "decoupling"), on the grounds of national security. This intensifies the fragmental, military and antagonistic tendency in cyberspace.

2. International mechanism in maintaining peace, security and stability in cyberspace is urgently expected

As new emerging countries come to rise in cyberspace, they play more part in drawing up international rules on cybersecurity. The present advancement in making international rules on cybersecurity neither satisfies the needs of emerging countries that hope to safeguard their own interests, nor accords with the new situation of cybersecurity. In the real political world, international institutional systems that took shape after World War II, including international security system, economic system, political system, communication system and technology & science system, are fundamentally transforming. International community, like government, industrial organization and corporation, have not forged consensus on the orientation and impact of order-centered transformation. Therefore, to cope with the strategic instability of international systems and set up the strategic stability of cyberspace in line with the information age present us a new challenge.

### 7.2.3 COVID-19 Intensifies the Oscillation of Cybersecurity

The leap in online service worsens the threats to cybersecurity. COVID-19 reduces human activities in physical space and enhances their communication & interaction in cyberspace, which objectively accelerates the turn of traditional offline business to online business and spurs the fast development of new business forms and models like cloud office, cloud health care and cloud education. Yet, protective measures on cybersecurity have not advanced and improved in a timely manner in cloud-end business. Remote online service management mechanism and remote-environment-based cybersecurity defense system remain vulnerable and open to hackers in cyberspace.

In COVID-19, malicious attacks crippled the strategic mutual trust among countries in cyberspace. COVID-19 raged globally and global conditions tensed. Under such circumstances, countries in the world ought to strengthen cooperation and jointly resist COVID-19; nevertheless, several countries provoked incidents on the COCID-19 and politicized COVID-19 issues, viewing it as a strategy in political games and suppressing others. They even wantonly started rumors and discredited other countries, groundlessly accusing other countries of launching cyberattack to steal vaccine research findings and fabricate the origin of COVID-19 virus. This tremendously impaired mutual trust among countries.

1. COVID-19 lashes global industrial pattern and affects cybersecurity

COVID-19 plays havoc with world cybersecurity and stability, foregrounding the trend of cyber protectionism. Many countries have issued various policies that aim to enhance the protection of national organizations from cyber threats.[1] Cybersecurity protectionism intensifies the game in cyberspace. COVID-19 also impels various countries to deeply examine the security of the global industrial chain system based on traditional division of labor, quickens the localization and regionalization of the global industrial chain, and increases the trade barrier of cyber-information technology, geopolitical tension and the uncertainty in international cooperation. To establish safe and resilient supply chains & industrial chains of cyber-information technological products poses a challenge for the information-oriented development of countries in the world after COVID-19.

2. International exchange and cooperation in cyberspace are hampered in the COVID-19 era

Under the impact of the COVID-19 prevention & control and the restricted cross-border flow of people, many international exchanges in the field of cybersecurity had to be postponed or even cancelled. Objectively, COVID-19 suspended security governance in cyberspace and significantly affected international cooperation on cybersecurity. Now, it is a critical time to establish rules and regulations on cyberspace. The absence of security-governance mechanism undermines the attack upon cybercrime and the effectiveness of cybersecurity protection.

[1] *Cyber Protectionism: Global Policies Are Adversely Impacting Cybersecurity* released by SANS. See https://www.sans.org/reading-room/whitepapers/riskmanagement/paper/39115.

## 7.3 World Cybersecurity Threats Show New Features

Affected by geopolitical factors and the COVD-19, world cybersecurity threats showed new features in 2020: APT attacks displayed more geopolitical color, supply chain security faced enormous challenges, phishing became excessively active in the outbreak of COVID-19, and health care industry received the biggest threat. In brief, as cloud business developed fast, cybersecurity faced both opportunities and challenges.

### *7.3.1 Trends of Threats in Attacks on Cybersecurity*

1. APT Attack

APT attack synchronizes with geopolitical competition. In recent years, APT attack with national background has become the most serious threat to the security of cyberspace. The organization and model of APT attack constantly alter, yet beneath the fog of APT attack lie the game and contest of national cyber strength. The implacable confrontation in geopolitics firstly induces APT attack at the national level. For a region where regional politics tenses and regional security becomes complex, APT attack proves the most serious, frequent and complex.

APT attack that targets industrial control systems is detrimental. Since 2019, APT attack has shifted from political, military and diplomatic purpose to industrial control systems and key infrastructure, with unpredictable effect and incalculable loss. Compared with simple political, military and diplomatic purpose, cyberattack on industrial control systems and key infrastructure easily initiates "avalanche effect" among citizens and society owing to the systematic interconnection and dysfunctional domino effect.

2. Malware

Web-oriented malware multiplies significantly. As Kaspersky's statistical data suggests, in its Web antivirus platform, 24.61 million pieces of malware were detected in 2019, up by 14% over 2018, and *c.* 20% of Internet users were attacked by these malware.[2] Besides, malware attacks via hacking tools prevailed in 2019, up by 224%.[3]

Corporation-oriented malware increases. According to Malware Bytes' statistical data, malware attack dropped by 2% on individual consumers yet grew by 13% on corporations. According to *2019 IT Security Economics* released by Kaspersky, malware infected around 50% of corporate devices and 50% of staff devices in 2019. With more skillful tactics, malware attack casts eyes on more profitable targets.

[2] Source: Kaspersky website, https://go.kaspersky.com/rs/802-IJN-240/images/KSB_2019_Statistics_EN.pdf.

[3] malwarebytes, https://resources.malwarebytes.com/files/2020/02/2020_State-of-Malware-Report.pdf.

3. Data Leakage

Data leakage happens frequently. In May 2020, Verizon released *2020 Data Breach Investigations Report* (DBIR), in which 81 organizations from 81 countries were involved in the investigation of 157,525 security cases. In specific, 32,002 cases met the research criteria, and 3,950 cases were confirmed as real data breaches. 81% of data breach were discovered within several days or more quickly, 43% of data breach was associated with Web application, and 58% of victims' personal data was disclosed.

Cybercrime becomes more complex and targeted. In December 2019, Qian'xin published *Analytical Report on Typical Cases of Data Leakage*, which revealed that presently, data leakage was featured with diverse and covert means, and that real-time data monitoring was challenged. In addition, the risk of data leakage mainly existed in internal staff (as high as 80%), who were motivated by obtaining business profits. Users' data was severely leaked, accounting for 35% of data leakage, followed by trade secret data and customer data. In the external personnel who disclosed data, 90% were hackers, who took advantage of vulnerabilities to decode high-authority accounts, directly edit the database, and steal data.[4] In *Analytical Report on Typical Cases of Data Leakage*, it was also revealed that business profits remained the most powerful motive for data stealing.

4. Ransomware Attack

In 2020, ransomware was a great threat to cybersecurity with enormous loss. According to *2019 IT Security Economics* released by Kaspersky, around 40% corporations it investigated encountered ransomware attack and suffered a loss of $1.46 million on average in each case. To curb ransomware virus, Europol and ICPO issued *2019 Internet Organised Crime Threat Assessment* on October 9, 2019, which stated that ransomware remained the biggest threat to cybersecurity.

The technological upgrading of ransomware attack makes it harder to detect and resist. Ransomware attackers launch attack more actively and successfully, combine their creativity with automatic tools, and maximize their attacks. APT organizations with national background collect intelligence and conduct sabotage by means of blackmail. Ransomware virus continually searches sensitive information in infected targets, repeatedly encrypts infected data, or even irreparably destroys files. Screened by ransomware virus, ransomware attackers ruin data, erase the trace of intrusion, and conceal the real intention of attack. "Man-aided virus" increases, by which attackers can pinpoint the targets and analyze the infected networks to identify the most critical systems in the network and obtain their passwords. After that, ransomware can be released to maximize the damage to data.

---

[4] Source: *Analytical Report on Typical Cases of Data Leakage*, December 2019. See https://www.qianxin.com/threat/reportdetail/37?type=report_apt_list.

5. DDoS Attack

DDoS attack features a large bandwidth and a high throughput rate. According to *2019 IT Security Economics* released by Kaspersky, it was indicated that in the first quarter of 2020, the number of high-capacity attack continued to increase, and the number of attack above 50 Gb/s reached 51. The average bandwidth of attack also rose, from 4.3 Gb/s in the first quarter of 2019 to 5.0 Gb/s. In *Global DDoS Threat Landscape Report* released by Imperva, a cybersecurity company, it was revealed that in 2019, network-level DDoS attack set a record of 580 million requests per second, and that in April 2019, application-level DDoS attack lasted for 13 days, with the peak throughput rate of 292,000 requests per second.

The effectiveness of DDoS attack continuously improves. In September 2019, Akamai reported a new DDoS vector, Web Services Dynamic Discovery (WSD), which is a multicast discovery protocol for local network location service. Helped by WSD, cyberattackers can extensively locate and destroy mis-configured networked devices, which enlarges the scope of DDoS attack. Experts from Kaspersky view that by virtue of the rise of 5G and the increase of IoT device, the traditional boundary of critical infrastructure like water supply, power grid, military facilities and financial institutions will extend to other new fields in 5G world. These all call for new security standards, and the rise of connecting speed will create favorable conditions for DDoS attack.

6. Phishing

Phishing integration tools spread widely. Akamai's *SOTI Report* claims that phishing integration tool developers sell "phishing service software". Some developers own shops and advertise them on social media. The price of integrated tools reaches $99 and rises with chosen mail attack service. All integration tools display security and elude detection. Phishing integration tool developers will offer more sophisticated products to further enhance the function in launching phishing activities. Cyber-attackers can improve the effectiveness of their phishing activities by reducing or concealing the common signs of phishing. If cyberattackers use deceptive ways like leaked internal or third-party accounts to send phishing attempts that seem legal, more corporate emails would be leaked.

In the COID-19, phishing was extremely active in cyberspace. In *Threat Spotlight Report* newly published by Barracuda, a cybersecurity company, it was exposed that email attacks on the COVID-19 virus continued to rise. On March 1–23, 2020, Barracuda totally detected 467,825 times of email attack, of which 9,116 concerned the COVID-19 virus and accounted for 2% of the total. Zscaler, a cybersecurity company, declared that more and more hackers lured victims by promising to provide the information or protection against the COVID-19 virus, with around 20,000 cases of phishing attack. In these attacks, users were directed to deceptive websites and inveigled into entering sensitive information. By March 23, 2020, phishing attack that centered on the COVID-19 virus could be categorized into three types, i.e. deception (54%), brand counterfeiting (34%) and email blackmail (11%).

7. Attacks from Vulnerability

The vulnerability of apps remains worrying. According to *Vol. 10 Survey Report on Software Security State* published by Veracode, a cybersecurity company, in 85,000 apps Veracode tested, 83% of them had at least one vulnerability in security and 20% of them had at least one serious vulnerability (more in many apps). In the test, there were 10 million vulnerabilities discovered in these apps. As a result, cyberattackers got access to many potential and available vulnerabilities, including "zero-day".

The security of open source software becomes grim. In March 2020, WhiteSource, a provider of security & license compliance management solution, released *2019 Open Source Components Security Vulnerability Status Report*. As the data suggests, in 2019, the number of vulnerabilities on open source software soared to more than 6,100, up by *c.* 50% over 2018, which resulted from the widespread use of open source components. Synopsys, a software and chip design company, asserted that unsafe open source software could be seen everywhere, in *2020 Open Source Security and Risk Analysis (OSSRA) Report*. On the one hand, 99% of audited code bases contain at least one open source component. On the other hand, 75% of audited code bases contain open source components with known security vulnerability, and 49% of audited code bases contain high-risk vulnerabilities, with aging and abandoned open source components widely seen. Presently, as the application of open source components into software continually grows and the security of components increasingly becomes serious, the discovery and management of open source (or third-party) components serves as one major or mandatory function of AST solutions.

Remarkable progress has been made in vulnerability patching. In late 2019, GitHub launched GitHub Security Lab, in which researchers and developers on cybersecurity patched vulnerabilities and shared expertise, aiming to improve the overall security of GitHub code-sharing ecosystem. In Verizon's *2020 Data Breach Investigations Report* (DBIR), it was conveyed that vulnerability patching proved better than expected, and that only 2.5% of SIEM (Security Information and Incident Management) incidents involved the use of vulnerability. DBIR proclaimed that "the finding implied that most organizations did quite well in vulnerability patching". Notwithstanding, vulnerability meant a disadvantage for cybersecurity. Besides, 70% of vulnerability attacks were caused by external hackers, and the biggest incentive was the pursuit of business profits. 86% of vulnerability attacks were induced by business reasons and 10% of vulnerability attacks by espionage.

8. Cryptocurrency Hijacking

Cryptocurrency attack gradually decreases. The incidence of cryptocurrency-oriented attack alters with the value of cryptocurrency. Experts from Kaspersky affirm that mining attack becomes obsolete and attack on home users declines under the influence of lower profits. In various countries, new regulation & coordination systems sustainably improve, the security level of cryptocurrency transaction continually advances, and cryptocurrency-oriented hacking attack in cyberspace gradually lessens. As a survey suggested, in 2019, cryptocurrency attack averagely caused a loss of $1.62 million to corporations.

9. Supply Chain-Oriented Cybersecurity "Stepping Attack"

Third-party providers have been targeted in cyberattacks. According to *2019 IT Security Economics* released by Kaspersky, in big companies and small & medium-sized companies, the similarity rates of cyberattack incidents related to third-party providers (service and product) reached 43% and 38% respectively. According to a survey by One Identity, 94% companies licensed third parties to visit their network, and 72% companies privileged third parties in visiting their network. Yet, only 22% companies expressed their confidence in allowing third parties to visit unlicensed information, and 18% companies reported data leakage due to the visit of third parties. Companies, suppliers and partners will be closely tied in a digital way; unluckily, cyberattackers become more skilled.

Hardware becomes a new target. Hackers now attach greater importance or shed more eyes on hardware. Under any circumstances, malicious intrusion into hardware stack (e.g. firmware, BIOS and UEFI) means an enormous threat, which is amplified in supply chains.

The security of supply chains encounters a great challenge. In the past few years, supply chains turned into a new stage of cybersecurity. This was evidently signaled by more speeches themed on hackers' intrusion into supply chains at BlackHat and Defcon. Recently, CrowdStrike completed a global survey of 1,300 companies, demonstrating that 90% companies stayed "unprepared" for supply-chain cyberattack. *2019 Global Cyber Threat Report* noticed that more than 50% of cyberattack now utilized "Stepping Attack", which signified that attackers targeted more than one organization. The latest reports of Accenture[5] and BSI[6] both viewed supply-chain cybersecurity as the biggest challenge today.

### *7.3.2 Trends of Threats to Major Platforms & Fields and Cybersecurity*

1. Security of cloud platforms becomes a hot field in cybersecurity in recent years

Cloud platforms have been targeted in data stealing. In July 2019, major global media all covered the largest-scale data leakage case in American financial industry that happened to Capital One, which engulfed around 100 million American citizens and around 6 million Canadian citizens. It was reported that attackers utilized misconfiguration and downloaded extremely sensitive data, causing a direct loss of $150 million. Afterwards, FBI disclosed that 30 organisations with similar misconfiguration were fatally attacked too. In April 2020, SOS, the "safest online backup service

[5] *2019 Global Cyber Threat Report* released by Accenture. See https://www.accenture.com/_acnmedia/PDF-107/Accenture-security-cyber.pdf#zoom=50.

[6] *2019 Supply Chain Risk Analysis Report* released by BSI. See https://www.bsigroup.com/globalassets/supplychain/localfiles/us/reports/bsi-screen-supply-chain-risk-insights-for-2019.pdf.

provider in the world", suffered massive data leakage. As vpnMentor, a research team, suggested, SOS Online Backup had leaked the personal records of more than 135 million online customers.

Loophole on internal management forms a major incentive to cloud-security cases. (ISC)$^2$, a professional organization on world cybersecurity, published *2019 Cloud Security Report*, stating that 28% companies met with cloud security cases, and that the biggest loophole lied in unauthorized access like the abuse of staff pass and loose control of illegal access, with the configuration error of cloud platforms accounting for 40% of public cloud security cases.[7] According to a report released by Gartner, in 2919, lots of data leakage cases resulted from the misconfiguration of cloud storage platforms, and 95% of cloud security cases came from customers' fault rather than cloud infrastructure providers.

2. The risk of IoT security emerges with the wide use of IoT

The risk in IoT turns into a biggest risk for companies. U.S. Security Industry Association (SIA) released *2019 Security Megatrends Survey Report*, which signaled that IoT and the data it generated greatly impacted on employees in security industry in 2019. For most companies, to cope with the threat from IoT was a cardinal task in 2019.

Vulnerabilities of IoT-device security remains severe. *2020 Global IoT/ICS Risk Report* released by CyberX listed the most common security vulnerabilities that made IoT devices subject to attack in the last 12 months. The report clarified that admittedly, some vulnerabilities were remarkably improved, e.g. a 30% down in the number of remote-access IoT devices, yet vulnerabilities were found in 54% IoT devices; that the number of vulnerability in directly-connected IoT devices dropped from 40 to 27%, yet outdated operating systems existed in 71% websites (53% in 2019); that 66% websites failed to update automatic antivirus devices (43% in 2019).

3. The threat and attack to industrial control system security continue to rise

Tool-aided attack technology intensifies the threat to industrial control system security. According to FireEye's latest *Industrial Control System Security Report*, as attack tools for industrial control systems sprang up, ordinary hackers could launch attacks that demanded particular knowledge and advanced hacker skills in the past. In the future, industrial control system security would face grimmer situations. The free-of-charge Metasploit Framework has less ICS-targeted vulnerability apps; however, these vulnerabilities form huge risks for defenders. Tools for industrial control system attack significantly lowers the threshold of attack, which means a biggest threat at present and in the future.

Industrial control systems must cope with serious risks for cybersecurity. Researchers from FireEye, a cybersecurity company, explained that hundreds of ICS-targeted vulnerability use-modules related to more than 500 vulnerabilities were tracked, and 71% of them were potential "zero-day". In February 2020, Dragos, an

[7] (ISC)$^2$, *2019 Cloud Security Report*, www.isc2.org.

industrial cybersecurity company, released *2019 ICS Vulnerability Situation Report*, stating that 77% vulnerabilities covered systems that generally existed in the deep part of control system networks, including engineering workstations, HMIs, operator panels, industrial network devices and field devices. More than 50% vulnerabilities probably expunged visibility (failing to monitor or read system state) or control (failing to modify system state). It is a hard task to discover vulnerability and use it to launch attack with real damage. However, taking into account the facts that attackers' TTPs (tactics, techniques and procedures) become complex and attackers' ability evolves quickly, and that industrial systems span a life cycle of more than 10 years, it means an unprecedented challenge for defenders to patch vulnerabilities faster than attackers who use vulnerabilities. As IoT devices multiplied and cybercrimes raged, the global industrial field faced serious cybersecurity risks in 2020.

## 7.4 Various Countries in the World Have Made New Achievements in Formulating Policies on Cybersecurity

Various countries in the world actively make a series of policies & measures on cybersecurity and consolidate the foundation for cybersecurity protection. Data security, personal-privacy protection and new technology remain major fields for formulating policies on cybersecurity in these countries.

### *7.4.1 Establishing Strategies on Cybersecurity: A Survey*

Many countries positively improved strategic plans and related action plans on cybersecurity, aiming to strengthen their cybersecurity-protection capabilities. In September 2019, Australia released *Australia's 2020 Cybersecurity Strategy* (draft for comments), which comprehensively reviewed many security programs it implemented since the release of *2016 Cyber Security Strategy*. Australia would also propose an assessment plan and clarify the detailed path for achieving strategic objectives. In October 2019, Poland launched *Polish Cybersecurity Strategy (2019–2024)*, which aimed to improve Poland's cyber "resilience" and better protect the data of public & private sectors and military systems. In February 2020, Brazil published *National Cybersecurity Strategy (2020–2023)*, which proposed the strategic vision of forging Brazil into a "cybersecurity nation" and established three strategic objectives on cybersecurity. European countries actively implemented relevant policies and measures. In March 2020, U.S. Cyberspace Solarium Commission released *Layered Cyber Deterrence Strategy*, which aimed to develop new cyber strategies for the United States. By 2019, more than 110 countries in the world developed or were developing cybersecurity strategies.

### *7.4.2 The Protection of Data Security Has Been Constantly Enhanced*

1. Data protection systems continuously improve

To enhance the protection of citizens' personal information, Brazil passed a constitutional amendment that incorporated the protection of personal data on digital platforms into Brazilian constitution as a fundamental right of citizens. In September 2019, in Singapore, newly revised *Personal Data Protection Regulations* came into force, which ruled that information like ID card number could only be requested from the public on the condition that Singaporean laws required or citizens' identity must be confirmed. In March 2020, Japan authorized an amendment to the *Act on the Protection of Personal Information*, which required companies to obtain users' consent before they provided personal data to third parties, e.g. Internet browsing history.

2. Cross-border data flow receives wide attention

The European Union maintains the existing cross-border data management strategy, *General Data Protection Regulation* (GDPR), which prioritizes privacy protection and emphasizes the cross-border flow of personal data. The United States underscores the free flow of data, yet strictly regulates and controls cross-border data when it comes to core interests. In November 2019, the United States enacted *National Security and Personal Data Protection Act (Draft)*, which highlighted the localization of data and prohibited data-flow overseas. In February 2020, the final rules of *Foreign Investment Review Act* of the Committee on Foreign Investment in the United States (CFIUS) was enforced, which strictly controlled the investment from foreign businessmen in the field of sensitive personal data to prevent the leakage of sensitive data. The Council of Minister of India approved *Personal Data Protection Act*, which stipulated that sensitive and important data must be stored and processed in India. Egypt passed *Data Protection Law*, which prohibited the transfer or sharing of personal data among foreign countries. In March 2020, Australia amended *Telecommunications (Interception and Access) Act* based on the *Clarifying Lawful Overseas Use of Data Act (Cloud Act)* to strengthen bilateral agreement and cooperation with the United States. It allowed agreement-signing countries to access communications data overseas for legal purpose.

3. Legislation in the field of data security displays a political trend

The United States and the European Union propose several laws on "secure election", which ban foreign citizens' participation in any election-related activities. The U.S. House of Representatives and the U.S. Senate have introduced or passed several proposals, like *Voter Privacy Act* and *Secure Elections Act*, to improve the election voting system of the federal government, intensify the supervision of online advertising, prohibit foreign citizens from interfering in election, and boost the protection of voters' personal privacy. The European Union has adopted legislative motions like *Resolution on Foreign Electoral Interference and Disinformation in National and European Democratic Processes* to protect the electoral security of its member states.

4. Law enforcement on personal information protection continually intensifies

Personal information protection standards in the field of subdivided technology have been successively formulated. The United Kingdom and France updated the guidelines for the use of Cookies stored on users' local terminals as well as the existing laws and regulations on privacy and electronic communications, stressed the fundamental role of "user consent" in the use of Cookies, and introduced cast-iron protection standards on personal data in terms of Cookies. Particularly, in COVID-19, various countries actively used digital technology to combat it. Personal-privacy protection centered on COVID-19-related information, with relevant protection remarkably strengthened. The United States proposed *2020 COVID-19 Consumer Data Protection Act* to regulate the collection and use of personal information related to the COVID-19 pandemic. The European Union published *Guidance on Apps Supporting the Fight against the COVID-19 Pandemic in Relation to Data Protection* to ensure that citizens' personal data would be adequately protected when they use apps. Japan Personal Information Protection Commission (PPC) released G*uidelines on Data Processing of Contact-Tracking Mobile Apps*, ordering "business operators of personal information processing" to strengthen the protection of personal information.

### 7.4.3 The Protection of Critical Information Infrastructure Continues to Be Strengthened

The United States enacted a series of protection-related bills and programs on critical infrastructure. Specifically, *Protecting Critical Infrastructure Against Drones and Emerging Threats Act* required U.S. Department of Homeland Security to evaluate related terrorist threats from drone technology. *The Securing Energy Infrastructure Act* (SEIA) planned to protect U.S. electric grid from hackers by replacing automatic systems with low-tech redundancy. U.S. Government Accountability Office published a reported entitled *Critical Infrastructure Protection: Actions Needed to Address Significant Cybersecurity Risks Facing the Electric Grid*, which analyzed cybersecurity risks and challenges that U.S. electric grid faced and delineated the actions that federal agencies took to address grid cybersecurity risks. *Layered Cyber Deterrence Strategy* of the United States prioritized the legislation of the idea of "infrastructure with systematical and critical significance".

### 7.4.4 Policies on Technological Field Have Been Actively Enacted

1. Various countries speed up the deployment of 5G security

To ensure the development of 5G security, various countries in the world have issued relevant strategies and policies. European Commission released *EU Coordinated Risk Assessment of the Cybersecurity of 5G Networks*. The United States approved several 5G security-related bills, including *National Strategy To Secure 5G of the United States of America* and *The Secure 5G and Beyond Act*. However, in 5G issues, the United States played up hegemonism, unilateralism and protectionism, politicized 5G technology issues, generalized the concept of "national security" and concocted the "theory of national origin". This had an enormous impact on the stability of the global 5G industrial chain and the security of supply chains, and posed a threat and challenge to the development of 5G security.

2. The development of AI security sustainably advances

The United States released *The National Artificial Intelligence Research and Development Strategic Plan: 2019 Update*, which emphasized building healthy and trustworthy AI systems to ensure the development of AI security. The European Union Agency for Network and Information Security (ENISA) published the report called *Towards a Framework for Policy Development in Cybersecurity: Security and Privacy Considerations in Autonomous Agents*, which aimed to provide EU member states with a policy-making framework to address security and privacy issues that resulted from AI. High-Level Dialogue on World Artificial Intelligence Security and the Rule of Law Forum of World Artificial Intelligence Conference jointly released *Guidelines on Artificial Intelligence Security and Law 2019*, which scientifically predicted the security risks of AI development from five aspects and proposed countermeasures to the risks, in order to protect the "security gene" of AI development. In June 2019, G20 Ministerial Meeting adopted *G20 AI Principles* to promote the establishment of national policies and international cooperation on reliable AI.

## 7.5 New Trends in the Technological Development of World Cybersecurity

Technological development of world cybersecurity advances in stability, featuring the wide application of zero trust architecture, the increasingly prominent role of human in cybersecurity, and more intelligent means available to solving cybersecurity issues.

### 7.5.1 *Zero Trust Architecture Multiplies*

Zero trust is a new approach on cybersecurity, meaning "No Trust & Forever Verification". Each device, user and network process must be verified, recorded and checked. Zero trust security architecture re-assesses and re-examines the idea on traditional boundary-security architecture, building a totally new identity-centered boundary.

In recent years, zero trust-based cybersecurity constituted an important part of cybersecurity policies and measures in some countries, after many-time trials and tests. Zero trust would turn into the first choice for cybersecurity architecture in the future. In February 2020, PulseSecure, a global zero trust security access company, publicized *2020 Zero Trust Report*.[8] After investigating more than 400 security decision-makers in various industries like finance, health care, manufacturing, high-tech, government and education, PulseSecure unlocked that 53% companies planned to transfer zero trust access function to hybrid IT deployment, and 25% companies would apply SaaS-based zero trust security framework.

### 7.5.2 *Human Factors Play a Prominent Role in Cybersecurity*

In recent years, cybersecurity affirmed the basic understanding of "human as crucial role" in practice. Cybersecurity preferably concentrates on the latest and most advanced technologies, yet often ignores the human factor. As a matter of fact, it is human that plays an everlasting role in cybersecurity protection and cyberattack. Human factor remains a weak link in cybersecurity management. Cybersecurity incidents generally arise from malicious or unintentional human errors. According to a research done by IBM, human errors induce 95% cybersecurity vulnerabilities. Besides, phishing, a major means of cyberattack, takes advantage of human's curiosity and carelessness.

### 7.5.3 *Intelligent Means Are Used to Solve Problems on Cybersecurity More Widely*

In cybersecurity, AI was firstly applied to simple use cases, e.g. screening junk mail for email users. Since 2020, AI technology has extended to all functions and departments of cybersecurity teams. In recent years, cybersecurity companies at home and abroad effectively used the powerful learning and reasoning capability of AI, and launched many AI-based tools or products for malicious code detection, abnormal traffic detection, software vulnerability mining, abnormal behavior analysis, sensitive data protection and security operation management. This effectively promoted the

[8] *2020 Zero Trust Report* released by PulseSecure, February 2020. See https://www.pulsesecure.net/resource/2020zero-trust-report/ .

accuracy and efficiency of cybersecurity defense. Advances in AI triggered security systems that were more intelligent and more autonomous. With the help of machine learning, these systems could learn and improve independently without human intervention. AI-based cybersecurity protection & application became a major direction in the field of cybersecurity. Practically, cybersecurity integrated with AI more widely, represented by big data analysis, machine learning, deep learning and man–machine collaboration.

## 7.6 Cybersecurity Industry Develops Sustainably

Various countries in the world continually improve cybersecurity policies and standard systems, and enhance investment in cybersecurity. This lays the foundation for the development of cybersecurity industry, which presents a trend of fast development.

### *7.6.1 World Cybersecurity Industrial Scale Grows Steadily*

In 2019, world cybersecurity industrial scale reached $121.7 billion, up by 8.64% year on year. The distribution pattern of world cybersecurity industry basically remained stable. North America, Western Europe and the Asia–Pacific Region stood comparison with each other, whose market share exceeded 90%.[9] In specific, North America, mainly the United States and Canada, achieved a market scale of $54.8 billion and accounted for 45% of global market, up by 9.6% over 2018, which took the lead in the world. Western Europe, mainly the United Kingdom and Germany, realized a market size of $31.6 billion and accounted for 26% of global market, up by 7.5% over 2018. The Asia–Pacific Region, mainly China, Japan, Australia and India, realized a market scale of $26.8 billion, and accounted for 22% of global market, up by 8.9% over 2018. In the Middle East, Eastern Europe, Latin America and other regions, the market scale of cybersecurity industry reached $8.5 billion and accounted for 7% of global market, up by 6.3% over 2018. See Fig. 7.1 for "The Regional Distribution of World Cybersecurity Industry in 2019".

### *7.6.2 Cybersecurity Industry Advances Progressively*

1. Cybersecurity service market keeps developing well

Cybersecurity service becomes a major field in the world cybersecurity market. Security service and security product account for 52% and 48% respectively. As defined

[9] Source: *2020 Global Cybersecurity Industry Market Analysis*, April 2020. See https://bg.qianzhan.com/trends/detail/506/200417-3d8590cc.html.

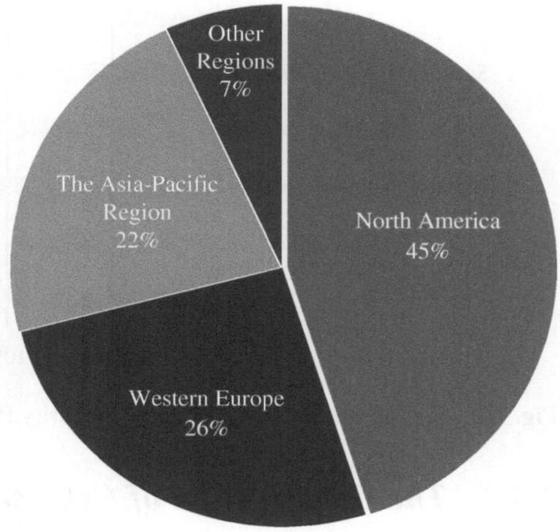


**Fig. 7.1** The Regional Distribution of World Cybersecurity Industry in 2019

by International Data Corporation, the cybersecurity service market comprises four submarkets, i.e. consulting service, integration service, IT education & training and managed security service (MSS). Noticeably, the MSS market was worth $64.2 billion in 2019, twice the amount invested in infrastructure protection and cybersecurity equipment.[10] MSS provides users with accurate and efficient security services and helps corporations to raise the level of security defense & response, turning into new demand in the global market. As the survey data of Internet Data Center suggests, when global users choose MSS providers, they consider such factors as round-the-clock professional support, threat intelligence capability, detection capability, response capability and security strategy & planning. See Fig. 7.2 for "The Market Share of Global IT Security Product and Service in 2019".[11]

2. Cybersecurity market in key fields develops quickly

AI empowers cybersecurity significantly. In 2019, more than 180 cases investment and financing occurred in AI and cybersecurity, accounting for 45% of the total investment in cybersecurity. AI cybersecurity market now has a value of $8.8 billion and will grow at a Compound Annual Growth Rate (CAGR) of 23% in the future.[12]

---

[10] *Global Cybersecurity: Situation, Figures and Statistics (2020)*, March 2020. See https://baijiahao.baidu.com/s?id=1661404390660,561,771&wfr=spider&for=pc, https://securityintelligence.com/articles/11-stats-on-ciso-spending-to-inform.

[11] IDC < Worldwide Semiannual Security Spending Guide 2018H2 > .

[12] AI breeds enormous cybersecurity market, June 2020. See https://mp.weixin.qq.com/s?src=11&timestamp=1,604,630,441&ver=2689&signature=4*aGv-joy4V9e8VUkw5jTrITWdyesWWYVpO-QaF2DlSHbW6QO1odYg-IcOhP1Sn6fONV*Q9TtIo1bK1-Fk3rgUBvR7tBAOvRjiQ*uLlXWdBpurkdcMCFTyLdjoKh2eks&new=1.

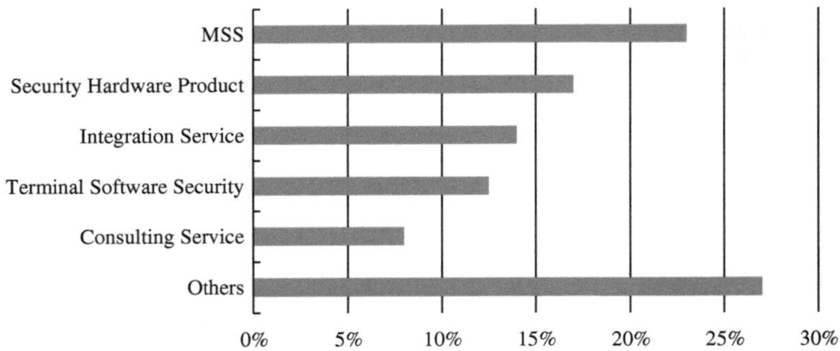


**Fig. 7.2** The Market Share of Global IT Security Product and Service in 2019

### *7.6.3 The Construction of Cybersecurity Industrial Ecosystem*

1. The scale of market demand continues to enlarge

According to *Cisco* 2020 *CISO Benchmark Report* released by Cisco in February 2020, 89% corporations gave priority to the development of cybersecurity and 61% corporations took the initiative in disclosing violative actions, showing the highest enthusiasm in the past four years. Major global corporations in the world enthusiastically raised the budget and investment on cybersecurity. According to *2020 CIO Status Survey* done by IDG, a global data consulting platform, 34% corporations invested the largest amount to security and risk management. Under the impact of COVID-19, teleworking became a normal way. The threat and risk on cybersecurity intensified, and the demand for cybersecurity protection rose simultaneously.

2. The investment and financing of corporations on cybersecurity prominently grow

Transnational capital shows a rising enthusiasm in cybersecurity industry. From January to November in 2019, the financing of world cybersecurity industry reached $7.322 billion, up by 18.1% over 2018.[13] Influenced by COVID-19, the market value of several cybersecurity corporations significantly changed. In March 2020, the market value of world cybersecurity corporations increased by $2.6 billion, or up by 10.2% over November 2019. Particularly, in the Asia–Pacific Region, the market value of cybersecurity corporations grew robustly, up by 41.9% on average. In other regions, the market value of cybersecurity corporations slipped, the United States down by 4.5% on average, Europe down by 8.7% on average, and Israel down by 18% on average (with a remarkable trend). See Fig. 7.3 for "The Change of Market Value of Major Cybersecurity Corporations in the World in November 2019 and March 2020".

[13] Source: *2020 Global Cybersecurity Industry Market Analysis*, April 2020, https://bg.qianzhan.com/trends/detail/506/200417-3d8590cc.html.

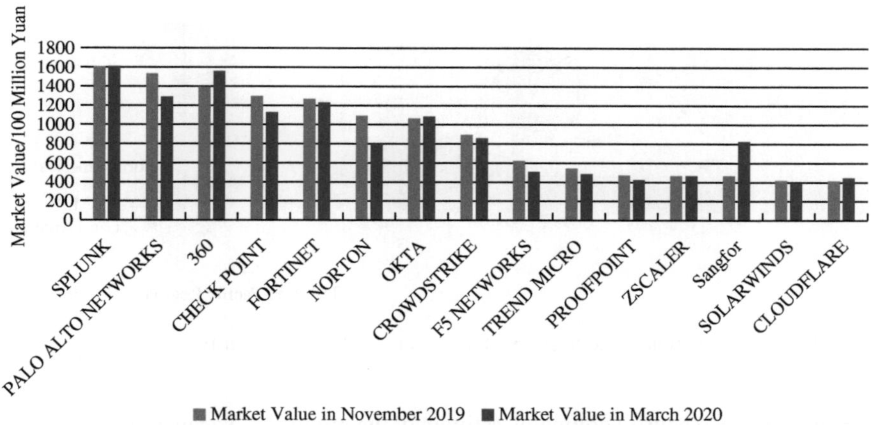


**Fig. 7.3** The Change of Market Value of Major Cybersecurity Corporations in the World

## 7.7 New Demands for Talent Training in Cybersecurity

As cybersecurity situation becomes complex and severe, and talents on cybersecurity remain insufficient, various countries view talent training in cybersecurity as a strategic means to promote national comprehensive cybersecurity capability, and take all-round measures to enhance talent training in cybersecurity.

### *7.7.1 The Lack of Talents and Technologies in World Cybersecurity Remains Grim Now*

Currently, the vacancy of talent on world cybersecurity becomes huge, and the demand for professional staff on cybersecurity continuously evolves, with the shortage of talent unsolved globally.

1. Large Talent Gap

Generally speaking, 70% countries in the world are nagged by the shortage of cybersecurity talent. (ISC)$^2$ *Cybersecurity Workforce Study* in 2019 announced that as many as 4.07 million workers in global cybersecurity were needed, which must be met with an increase of 145% in cybersecurity talent in the world.[14] 65% companies reported a shortage of cybersecurity talent, and 36% respondents expressed their deepest concern over the lack of skilled cybersecurity experts. As Cybersecurity

[14] (ISC)$^2$ *Cybersecurity Workforce Study*, 2019, Strategies for Building and Growing Strong Cybersecurity Teams.

**Fig. 7.4** The Distribution of the Vacancy of Talent on World Cybersecurity

Ventures, a cybersecurity company, estimated, the number of vacancy on cybersecurity would probably grow by 350%, rising from one million in 2013 to 3.5 million in 2021.[15]

2. Uneven Regional Distribution

In terms of cybersecurity talent, there are regional differences. The Americas and Europe own advanced Internet technologies and perform better than the Asia–Pacific Region and Africa or less developed regions. In 2020, the number of cybersecurity vacancy in Europe doubled to *c.* 300,000, which reached 560,000 in North America, 2.6 million in Asia (accounting for 64% of the world), and 600,000 in Latin America. See Fig. 7.4 for "The Distribution of the Vacancy of Talent on World Cybersecurity".

According to U.S. *Executive Order on America's Cybersecurity Workforce*, there are now more than 300,000 cybersecurity job vacancies in the United States, which accounts for 32%-45% of all technological job vacancies. According to the data released by (ISC)$^2$ in 2019, an estimated number of 560,000 workers must be added to fill the gap and meet the need for cybersecurity talent in the future, with a growth rate of 62%. Deloitte's research indicated that in 2021, Canada had the duty to supply around 8,000 cybersecurity jobs as expected.[16] A survey of the United Kingdom National Security Strategy Joint Commission revealed that the National Cyber Security Centre (NCSC) was challenged by the shortage of cybersecurity talent. In Ireland, technical deficiency in cybersecurity continued to worsen, and the demand for cybersecurity talent increased year by year, which required a growth rate of *c.* 18%. India National Association of Software and Service Companies (NASSCOM) estimated that in 2020, India would need one million cybersecurity professionals to keep pace with the rapid economic growth. India also had the largest number of cybersecurity job vacancies in recent years.

[15] Cybersecurity Ventures, The 2019/2020 Official Annual Cybersecurity Jobs Report. https://www.herjavecgroup.com/wp-content/uploads/2019/10/HG-CV-2019-Cybersecurity-Jobs-Report.pdf

[16] Deloitte, The changing faces of cybersecurity. https://www2.deloitte.com/content/dam/Deloitte/ca/Documents/risk/ca-cyber-talent-campaign-report-pov-aoda-en.pdf.

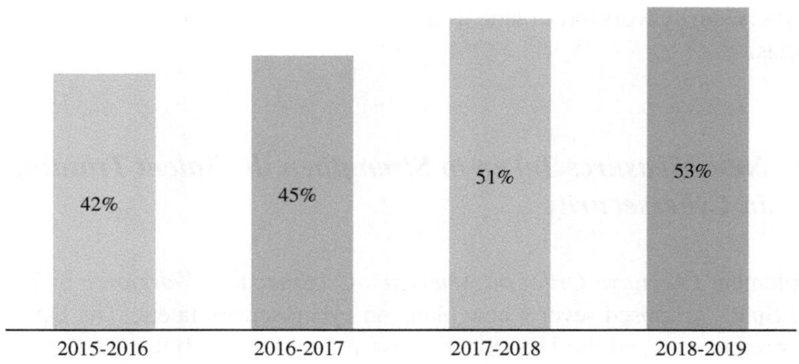


**Fig. 7.5** The Annual Percentage of Companies with Cybersecurity Skill Shortage in the Survey by Enterprise Strategy Group (ESG)

3. Inadequate Cybersecurity Skill

Stott and May, a cybersecurity research corporation, released *Cyber Security in Focus 2020*, noting that most cybersecurity leaders were coping with inadequacy in skill. 76% respondents observed that their organizations lacked cybersecurity skills, which marked a decrease by 12% over 2019. Around 72% organizations strived to attract cybersecurity talents, which remained stable over 2019.[17] Dana Foundation partook in the research. Its CISO Jim Rutt remarked that "Remarkably, outstanding talents like security architects, security engineers and senior security resolution specialists were needed". As relevant data suggested, the number of organisations that reported a serious shortage of cybersecurity skills kept rising since 2015. In 2018–2019, 53% respondents stated the shortage of cybersecurity skills.[18] See Fig. 7.5 for "The Annual Percentage of Companies with Cybersecurity Skill Shortage in the Survey by Enterprise Strategy Group (ESG)".

4. The diversity of talents in cybersecurity industry calls for attention

In terms of gender, the number of male talents substantially exceeds that of female counterparts in cybersecurity industry. As the data of Enterprise Strategy Group (ESG) implied, globally, only 11% women engaged in cybersecurity, which dropped to 8% in the United Kingdom. The data from $(ISC)^2$ declared that men outnumbered women among cybersecurity professionals in the world by more than two to one, with the highest percentage of females in Latin America (39%) and North America (34%). According to a report that the United Kingdom released in March 2020,[19]

---

[17] The Stott and May, Cyber Security in Focus 2020, https://resources.stottandmay.com/hubfs/Research/Cyber%20Security%20in%20Focus%202020_web-2.pdf.

[18] Jon Oltsik, The Cybersecurity Skills Shortage Is Getting Worse, https://www.esg-globalcom/blog/the-cybersecurity-skills-shortage-is-getting-worse.

[19] Source: Cyber Security Skills in the UK Labour Market 2020 released by UK DCMS, March 12, 2020. See https://www.gov.uk/government/publications/cyber-securityskills-in-the-uk-labour-market-2020/cyber-security-skills-in-the-uk-labour-market-2020.

15% cybersecurity workforce came from women, which reached 28% in other digital industries.

### 7.7.2 New Measures Taken to Strengthen the Talent Training in Cybersecurity

To implement *Executive Order on America's Cybersecurity Workforce* in 2019, the United States proposed several new plans on cybersecurity talent. The U.S. House of Representatives and the U.S. Senate passed a number of bills like *Cyber Talent Preparation Program Act*, which required U.S. Department of Labor to initiate an internal subsidy program to remedy the lack of cybersecurity talent and skill training. In February 2020, National Institute of Standards and Technology (NIST) released *A Roadmap for Successful Regional Alliances and Multistakeholder Partnerships to Build the Cybersecurity Workforce*, which summarized five NIST-funded cybersecurity talent training pilot projects, recorded the method and the best effect in practice, and provided a roadmap for building similar communities. In December 2019, the European Union Agency for Network and Information Security (ENISA) published a report, *Cybersecurity Skills Development in the EU*, which proposed to re-design education & training methods so as to improve the problems in the existing cybersecurity education, re-define the knowledge and skill on cybersecurity talent, and fundamentally transform the present condition of cybersecurity skill shortages.[20] In *Russia 2030: Science and Technology Foresight*, it was specified that in 2019–2030, Russian government would totally invest around $159.6 billion into forging a stable and efficient supporting system for scientific, engineering and entrepreneurial talents, and actualize a series of federal projects for talent training in information security.

[20] ENISA, Cybersecurity Skills Development in the EU, December 2019. https://www.enisa.europa.eu/publications/the-status-of-cyber-security-education-in-the-european-union/at_download/fullReport.

# Chapter 8
# The Construction of World Cyberlaws

## 8.1 Outline

In 2020, the construction of cyberlaws kept advancing, which improved the systematization and subdivision of the construction of cyberlaws. The innovation and application in Internet technology industry played an important part in the competition of comprehensive national power among countries. In the era of Internet, with the help of policies and legislative activities, how to promote the development of Internet industry, to maintain the security of Internet application and to balance the interests of multi-subjects received wide attention in the construction of cyberlaws recently. Various countries continued to deepen the legislation in the field of cybersecurity, grope for new approaches for the construction of cyberlaws, and improve the construction of the subdivided fields in cyberlaws, which were mainly fulfilled in four aspects:

(1) Emphasizing the protection of personal information and improving the rules of data protection in use cases.
(2) Regulating the order of Internet competition and purifying the content on the Internet.
(3) Enhancing the construction of cybersecurity mechanism and improving the system of operation guarantee.
(4) Ameliorating the development of new technologies and industries, and fending off the ensuing risks.

In COVID-19, the protection and utilization of personal information attracted enormous attention. Many countries took use cases into consideration and set up exceptional circumstances for the framework of the protection of personal information, attempting to balance the protection of personal information and other interests. Numerous problems arose from the competition among cyberspace platforms. The most prominent ones were monopoly and false information in cyberspace. Many countries regulated the competition among cyberspace platforms by legislative



Chinese Academy of Cyberspace Studies, *World Internet Development Report 2020*,
https://doi.org/10.1007/978-981-16-9388-5_8

and judicatory means in order to create a favorable environment for cyberlaws. As network technology developed, national competition expanded at the network level. Cybersecurity techniques & standards signaled a new direction in national competition. Various countries took multiple measures to strengthen the protection of cybersecurity and vied for leading position in new-era national competition. New industries represented by AI and IoT continually developed; meanwhile, the risks from their applications became a matter of great concern. Numerous countries commonly took cautious attitudes in supervision and legislation to ensure that all kinds of new technologies and new business forms developed in an orderly way in accordance with basic security standards.

## 8.2 The Legislative Work of Personal-Information Protection Further Advances, with the Formulation of Special Rules for Industrial Data Explored

In 2020, the legislative work on the protection of personal information deeply advanced. On the basis of establishing national "basic laws" on the protection of personal information, various countries started to explore special rules for the protection of personal information, according to the differences in data type and industrial feature & segment. The outbreak of COVID-19 incurred the turn in the central legislative work on the protection of personal information globally. Governments sought special rules for the collection and utilization of medical and health data, and reduced the institutional obstacles that governments and medical institutions probably encountered while evaluating and predicting the trend of COVID-19 as well as formulating policies on the prevention & control of COVID-19. In this way, governments fostered the free flow and timely sharing of epidemic data and medical & health information among governmental departments, medical institutions and research institutions.

### *8.2.1 Legislative Object Develops Deeply, with Specific Rules Continually Improved*

Many countries uphold the principle and orientation of personal information protection and specify the rules. Simultaneously, on the basis of the original principles of personal information protection, they consider use cases and formulate special rules for personal information protection.

In the protection of personal information, the United States adheres to the model that equally stresses personal information and commercial innovation. Yet, when they are set against each other, the priority is given to the intactness of commercial interests. For federal governments, bills are rarely approved by the U.S. House of

Representatives and the U.S. Senate. For state governments, 12 states like Nevada and California successively promulgate legislative files on the protection of personal privacy. For instance, in October 2019, the government of the State of California issued five amendments, including *California Consumer Privacy Act* (CCPA). As the legal texts issued by various states suggest, the basic concept that prioritizes commercial interests in the protection of personal information in the United States continues, placing stress on acquiescing in corporate collection of users' personal information and lowering the compliance cost of Internet companies. However, there are changes in sections. While strengthening the protection of sensitive personal information, the crux of legislative work turns from the original rules on personal-information deletion and disclosure to rules on personal-information collection and storage, which ensures that users' personal information will not be illegally abused by Internet giants after registering an account.

The European Union establishes the framework for individual rights, legal basis for data processing, accountability, data leakage report mechanism and law enforcement mechanism, which improves the feasibility in reacting to digital economy and scientific & technological innovation. Practically, the EU's *General Data Protection Regulation* (GDPR) stays in a predicament in which the flow and commercial use of personal data are restricted. EU lawmakers prefer to supplement and improve the EU framework for the protection of personal information by publishing detailed rules and guidelines in practice. In November 2019, European Data Protection Board (EDPB) issued the guideline that applied to Article 25 of GDPR, i.e. "Data System Protection and Default Protection". It was stipulated that legislators did not choose specific technologies or measures in security & application as compliance criteria, but judge in line with the practicality and effectiveness of personal data protection. Besides, data controllers only needed to meet the demand that technological means and security measures they took conformed to the protection of personal data in business and led to desirable legal effect. In the meantime, European Data Protection Board (EDPB) released *Guidelines on the Territorial Scope of the GDPR*, which clarified that GDPR also applied to data controllers or processors who set up business places and engaged in data processing within the territory of the EU. This broke the original restrictions of territoriality principle and nationality principle, and directly impacted on corporations that set up business places in the EU or engaged in data processing that involved industries and services in the EU. In terms of EU member states, in November 2019, Agencia Española de Protección de Datos (AEPD) passed two normative documents, i.e. *Guidelines for the Educational and Health Application of Personal Data* and *Guidelines for the Use of Cookies*, which further divided "Rules for Information Transparency" and "Statutory Forms of Consent" for data controllers in their collection of personal data. In particular, "clear-cut consent" from users must be obtained in order to use Cookie technology.

In their legislative work in 2020, Japan and South Korea also promptly adjusted the basic framework for the protection of personal information based on the status quo of cybersecurity, which continued to deepen the legislative work of personal information protection. In legislative trend, Japan supervised cross-border data processing, whereas South Korea chose to further develop basic ideas (e.g. personal information) connotatively and denotatively to strengthen the feasibility for the legislation of personal information protection. In March 2020, Japanese Cabinet approved *Amended Act on the Protection of Personal Information* (APPI). The amendment added legal obligations like reporting data leakage to Personal Information Protection Commission (PPC). Additionally, PPC's regulatory authority was extended to offshore companies, which was authorized to impose administrative fines on offshore companies involved in illegal data processing. In January 2020, National Assembly of the Republic of Korea passed three major amendments on data privacy laws, i.e. *Personal Information Protection Act* (PIPA), *Network Act* and *Credit Information Act*. These acts connected the protection of personal information with the developmental strategy of Korean digital economy, re-sorted provisions on personal information protection among various criteria with the help of pseudonymized information, and avoided the waste of regulatory resources caused by repeated rules.

### 8.2.2 Basic Principles for Information Protection Maintain Intact, with Exceptional Cases Allowed for Epidemic-Related Data

Under the impact of COVID-19, organizations on data protection in many countries commonly re-adjust the basic role of personal information protection, prioritize the free and safe flow of personal medical data, and stipulate several exceptional cases for the collection, use and flow of personal medical data under normal circumstances.

1. Determining the Basic Demands for Personal Information Protection

When a patient's health and medical data is processed, priority must be given to the protection of personal privacy and the right of health, the cardinal rule of the protection of personal information in COVID-19 must be clarified, and protective rules on special types of data in the legislation must be obeyed in the collection and use of health data. In March 2020, European Data Protection Board (EDPB) specifically promulgated *Guidelines for the Use of Health Data for Scientific Research under the Background of the COVID-19 Pandemic*, demanding that research-orientated health data processing must comply with EU's rules on basic rights, including consent requirements for personal data processing, principles of personal data protection, rights of data subjects and rules on cross-border data flow. Italian Data Protection Agency restated the legality of personal data processing. In specific, personal privacy must be protected in personal data processing; the principle of minimum processing must be strictly obeyed, with the processing of two types of sensitive information (health data and track data) emphasized and regulated. Besides, employers were forbidden to collect and process special information in a systematic and universal way, which violated laws and regulations. Data protection authorities supported public administrative departments that required their staff to undertake the responsibility of reporting information. The protection of personal information in the workplace was also stressed. In April 2020, Bundesamt für Sicherheit in der Informationstechnik (BSI, Germany's Federal Office for Information Security) released *Security Requirements for Digital Medical Applications* (BSI TR-03161), which clarified the security requirement for processing sensitive personal data in mobile health application.

2. Standardizing Rules for the Collection and Use of Personal Information

Administrative agencies and public organizations are allowed to collect citizens' personal information in line with special statutory authority, only for the purpose of preventing COVID-19 infection. For instance, UK Information Commissioner's Office (ICO) expressed that data protection law and electronic communications law would not inhibit government, NHS or any other health professional staff from sending public health information to citizens through telephones, short messages or emails. The Commission Nationale de l'Informatique et des Libertés (CNIL), a French data protection regulator, conveyed that health institutes could collect personal health data and had the right to take measures applicable to specific circumstances. Office of the Privacy Commissioner of Canada (OPC) specifically formulated rules for the collection of personal health and medical data in COVID-19 and clarified the cases for exceptional measures in *Privacy Impact Assessment Framework for the COVID-19 Outbreak Response*. Parliament of Australia passed *2020 Amendment to Privacy Act (Public Health Contact Information)*, which raised strict requirements for the collection, use and disclosure of user data done on COVID-19-related apps. The Act strived to ensure that data collected by COVID Safe (an app) was only used to assist Australia State and Territory health officials or serve their staff in tracking contacts. Besides, the data must help to maintain the normal operation of COVID Safe and national COVID Safe database, which accorded with the principle of integrity and security.

3. Removing Barriers to Communicate Information on COVID-19

Some countries allow organizations that obtain citizens' personal health data to share them with designated governmental departments for the purpose of protecting public interests. This aims to remove barriers to communicate information on COVID-19 among governmental departments, medical institutions and public organizations. UK Information Commissioner's Office (ICO) indicated that organizations could share staff's health information with governmental agencies for the welfare of public health. In a report it released in 2020, Agencia Española de Protección de Datos (AEPD, the Spanish Data Protection Agency) asserted that in the exigencies of health, the processing of personal data must conform to relevant regulations like GDPR as usual, in order to protect related personal rights. When the laws were enacted, the possibility of exceptional cases was predicted and the relevant principles of processing personal data were established, including legality, security, transparency, purpose limitation, accuracy and data minimization.

### *8.2.3 Data Leakage Occurs Frequently, with Administrative Penalty Tightened*

Governments in various countries watch out for the data collection and processing of Internet giants and make efforts to strengthen the administrative supervision on the legal criterion like purpose and necessity of data collection. For ordinary data controllers, governmental supervision on personal information protection centers on whether security vulnerability is caused by data leakage, whether corporate data security responsibility is implemented, and whether "informed-consent" rule is actually fulfilled.

In October 2019, as the large-scale incident of citizen-data leakage happened, Turkish government imposed a fine of $280,000 on data controllers who "failed to take necessary technological measures and administrative strategies" and "failed to fulfill the responsibility of data protection". In early 2020, as Google could not give users the right to delete, Swedish Data Protection Authority (DPA) confirmed that Google infringed the provisions of GDPR like "data subject rights", imposing a fine of around 7 million euros on Google. In early 2020, Facebook was sued and fined by several data protection agencies for data-leakage incidents. For example, in March 2020, Australian privacy supervision agency sued Facebook who undertook legal duty for data leakage. Brazilian Ministry of Justice reckoned that Facebook wrongly leaked the data of 443,000 Facebook users to individuals and ruled a fine of 6.6 million Brazilian Reais.

In the practice of personal information protection, U.S. Federal Trade Commission functions as a major regulatory body, whose essential regulatory standard lies in whether the data collection and processing of commercial institutions infringes the lawful rights and interests of consumers (users) and whether it impairs sensitive personal information. Retina-X and Johns, American Internet corporations, developed apps on mobile devices, which kept watch on users' mobile devices without telling them or obtaining their permission. In October 2019, U.S. Federal Trade Commission charged them with violating rules that banned unfair and deceptive practices, as well as *Children's Online Privacy Protection Act*. InfoTrax Systems, a tech startup in Utah, was sued by the U.S. Federal Trade Commission because it failed to take reasonable measures on security protection, which enabled hackers to obtain substantial customers' personal information. On November 15, 2019, the two parties were reconciled. Additionally, in 2019–2020, U.S. Federal Trade Commission was reconciled with Medable and T&M (New York) with regard to the use of privacy shield between the United States and the European Union.

## 8.3 Strengthening the Regulation of Cyber Platforms and Purifying the Environment of Cyber Platforms

Activities like data-resource monopolies and technological advantages elbowing out contracts constantly raise higher requirements for Internet market admittance. The United States and European countries start to take the business of Internet giants as a core link in cyberspace governance and conduct anti-trust investigation on activities that will harm or have harmed the competition order in Internet market. Meanwhile, the legislation on how to handle false information and purify the environment of cyber platforms has been put into agenda of Internet cyberlaws in various countries.

### *8.3.1 Standardizing the Competitive Means in Cyber Platforms and Maintaining the Free-Competition Order in Cyber Platforms*

Cyber platforms are an interdependent field of cyber technology and industrial development. To standardize cyber platforms helps to ensure the healthy and orderly development of cyber technology and industry. In July 2019, EU regulators announced an investigation into whether Amazon's use of sensitive data collected from independent retailers on its platform infringed EU's competition rules. The reason lied in that Amazon continuously collected sales data from retailers on its platform when it provided them with sales market, or that Amazon's collection of sales data aroused great concern of European Commission. In September 2019, the Commercial Court of Paris (Tribunal de Commerce de Paris) in France ruled that in terms of the contract between Amazon and the sellers on its platform, market dominant position was abused, with a fine of 4 million euros. In Amazon's contracts signed with sellers on the platform, Amazon had the right to alter the terms of the contracts at any time, like limiting delivery time or blocking delivery. In its survey report, French Directorate General of Competition, Consumption and Fraud Repression (DGCCRF) confirmed that Amazon's business activities above-stated constituted the "abuse of market dominant position", because in the contracts between Amazon and its suppliers, it was stipulated that "Amazon reserves the right to modify the contract and user policy at any time, without prior notice". This was identified as "obviously unfair" by DGCCRF that required Amazon to amend some terms with 180 days. Otherwise, Amazon would face a fine of 10,000 euros per day.

EU's supervision of super Internet platforms mainly centers on the supervision of personal data use and anti-unfair competition, keeping watch on the monopoly status of industrial platforms like Amazon and YouTube. In April 2019, European Commission published *Competition Policy for the Digital Era*, which revealed that owing to the uniqueness of market, only some cyber platforms survived in the market. Therefore, European Commission was on guard against that cyber platforms restricted industrial market admittance or imperiled neighboring market by using market dominant position. The rise of cyber platform giants remarkably affects the free competition order on the platform. For example, the most common means that cyber platforms take to restrict competition is to forbid sellers on the platforms to sell goods at a lower price on other platforms or via other channels, which constitutes a substantial obstacle to consumers' right to choose. In August 2019, 23 European recruitment websites asked EU Competition Commission to investigate Google's unfair competition, because Google's job software helped job-seekers to collect, screen and save various recruitment information and reminded them of relevant job opening hours. Besides, the software placed Google's job search results on the top, ahead of other job websites, which as EU regulators viewed, destroyed the free competition order in other markets. In the supervision of cyber platforms, the European Union always aims at EU consumers' rights and interests and cautiously regulates specific business of platform information service.

### *8.3.2 Protecting the Space for Consumers' Choice and Clarifying the Rules for Releasing Information on Cyber Platforms*

In July 2019, U.S. Department of Justice released *Justice Department Reviewing the Practices of Market-Leading Online Platforms*, stating that U.S. Department of Justice investigated commercial practice that harmed full market competition and users' interests, and revealed such commercial activities as how these giant platforms gained market dominance, reduced competition, strangled innovation or harmed consumers' rights and interests. U.S. Department of Commerce also investigated Internet giant platforms to assess the competition in cyber platform market in an objective and fair way and ensure that American citizens and corporations could enter free-competition market. In October 2019, 47 attorneys general from various states planned to participate in an anti-trust investigation led by New York State against Facebook, observing that Facebook could put consumers' data at risk, reduce consumers' choice and indirectly raise advertising price. In addition to Facebook, Google, Amazon and Apple were also considered potential monopolists. The investigation against Facebook touched upon all the business of these four corporations. Howbeit, owing to the complexity of American anti-trust law and the innovation of commercial models, American regulatory authorities could not directly judge that monopoly emerged as in traditional industries, fearing that these four technological corporations forged excessive "power" with their market advantage and user base, and restricted free competition and market admittance of other industries.

In April 2020, France Competition Administration Authority, a French anti-trust regulator, ruled that Google paid fees of news content to French publishing companies and news agencies, because Google recklessly released news and information of the former for free and published news or paper extracts in the link of its search architecture, which constituted the abuse of market dominant position. Legally, the above adjudication rested on *European Copyright Directive*, EU's latest rule. Prior to this, EU did not decide the belonging of the benefit of information collected by news aggregation platforms, yet as Article 11 of *European Copyright Directive* stipulated, paying licensing fees would benefit content producers, which forced Google to screen the thumbnail and textual information in the search results.

### *8.3.3 Enhancing the Supervision of Cyber Content and Purifying the Environment of Cyber Platforms*

As network technology develops, network becomes a new platform for human exchange and communication. It diversifies the way of human's information transmission, quickens information transmission, and further promotes the exchange and development of human thoughts. However, in an anonymous state, the publication of network information can hardly be controlled in an effective way, and cleaning up false information is used as an important means to purify cyber platforms.

In October 2019, Singapore's *Protection from Online Falsehoods and Manipulation Act* took effect. The Act gave Singaporean government the right to deal with online false information and provided legal basis by which government could require individuals or online platforms to correct or delete false information. It particularized the measures taken against false information and specified the measures taken against those who published and disseminated false information. Besides, the Act proposed that government could order Internet & mass media service providers to correct and block access to relevant false information, and stipulated that they undertook legal liability for false information activities, e.g. criminal penalty in some cases. In content, Singapore's *Protection from Online Falsehoods and Manipulation Act* provided comprehensive measures for the governance of false information. On June 2, 2020, European Commission initiated an open public consultation on *Digital Services Act*, which planned to strengthen the responsibility of online platforms in content supervision and urge EU member states to strictly supervise fake news and online rumors. On June 30, 2020, Brazilian Senate passed *Bill on Fake News* to curb and combat fake news and ensure the transparency of social media. With the text of the bill approved, senators voted on the amendment to the bill. Angelo Coronel, a senator, remarked that the legislation strived to simplify the text, actualize the right to obtain prompt reply and ensure the timely deletion of relevant content in serious cases. In terms of global trend, the regulatory and legislative course against false news accelerates, and many countries tighten their grip on the content supervision on online platforms, with the contention for national discourse power in cyberspace increasingly highlighted.

## 8.4 Promoting the Legislative Course of Cybersecurity and Improving the Guarantee System of Cybersecurity

In 2020, global cybersecurity faced grim situations, and cybersecurity incidents like data leakage and high-risk vulnerability frequently occurred. Various countries increased investment to strengthen their protective measures on cybersecurity. In major countries, special departments on cybersecurity were gradually established and public–private cooperation was progressively promoted. Meanwhile, major countries continually vied for a leading position in the development of 5G network and improved their operation & security systems.

### *8.4.1 Establishing Cybersecurity Departments and Admitting Social Participants*

The interconnection of network nodes determines that government-dominated cybersecurity construction proves fruitless and that a public–private-cooperated security guarantee system becomes urgent. Therefore, various countries have gradually established and improved public–private cooperation mechanisms on cybersecurity, and subdivided special functions of departments on cybersecurity. In terms of global landscape, U.S. federal government and state governments gradually strengthen governmental departments' control over cyberspace and expand the scope of administrative supervision. European countries prefer to enhance military force of cybersecurity, and the competition among them for cyber military force probably sees a new upsurge.

The United States establishes special cybersecurity administration departments and encourages social participation, which displays an obvious trend. In the second half of 2019, the U.S. House of Representatives National Security Council passed *Cybersecurity Advisory Committee Authorization Act*, which aimed to established a new cybersecurity advisory committee within Cybersecurity and Infrastructure Security Agency (CISA) under the command of the United States Department of Homeland Security (DHS), in order to comprehensively analyze cybersecurity threats of all industries. U.S. National Security Agency announced the establishment of the new cybersecurity agency and cybersecurity directorate to protect internal organizations from cyberattacks abroad. The United States continuously boosted the construction of special cybersecurity agencies in the national defense and military field. For instance, U.S. Navy set up a new CISO Office in charge of cybersecurity and data strategy. Moreover, the United States emphasized the cooperation with private organizations in cybersecurity, respected "whitehat hackers" who played an assisting role in cybersecurity defense systems, and built a dual protective mechanism for the public and private sectors. Additionally, various states strengthened the enforcement authority of cybersecurity regulators and the unitized institution of cybersecurity military force. In June 2019, for example, the State of Arkansas announced that it founded Arkansas Computer Science and Cybersecurity Task Force, which aimed to assess its cybersecurity education program.

Other countries also played an active role in promoting the construction of cybersecurity guarantee mechanisms. Russian Federal Security Bureau (FSB) enlarged the authority of Computer Accident Coordination Center, its subordinate agency. With the help of Domain Name Coordination Center, websites with cybersecurity threats were directly closed down, in order to ensure government's basic control over the operation of cyberspace. Poland planned to organize cyber defence force in 2024, which consisted of soldiers with cybersecurity qualification and experts with cybersecurity technology. The United Kingdom planned to establish National Cyber Force (NCF), which was co-founded by Government Communications Headquarters (GCHQ) and Ministry of Defence (MoD). NCF recruited hackers to retaliate against cybersecurity threats such as large-scale illegal access to databases, the dissemination of false news and cyber terrorists.

### *8.4.2 Improving Cybersecurity Systems and Attaching Great Attention to Supply-Chain Security*

The construction of network infrastructure plays a vital role in cyber era. Particularly, the innovation and development of network critical infrastructure represented by 5G technology triggers a global "reserve war" of network strategic resources, with cybersecurity as new content of national security. On the basis of cybersecurity basic laws and cybersecurity strategies, various countries gradually optimize the legislative work, and attach great attention to cyber sovereignty, vulnerability patching and supply-chain security in the construction of cyberlaws.

In 2019–2020, the U.S. federal government and state governments successively initiated cybersecurity laws, technological standards and national strategies. In September 2019, the U.S. House of Representatives passed *Cybersecurity Vulnerability Remediation Act*, and the U.S. House of Representatives and the U.S. Senate passed *Intelligence Authorization Act*. The latter required U.S. government to establish a working group on supply chain and counter-espionage risk management and take tracking cybersecurity threats as governmental long-term work, like working reports on 5G cybersecurity threats. At the state level, the State of New York passed *Stop Hacks and Improve Electronic Data Security Act*, which extended the reporting duty of data leakage, demanding that all legal entities that held business in the State of New York and owned personal information and data must undertake the duty of cybersecurity. In November 2019, U.S. Department of Commerce published *Executive Order on Securing the Information and Communications Technology and Services Supply Chain Implementation Act* and *Executive Order on Securing the Information and Communications Technology and Services Supply Chain*. The former gave U.S. Secretary of Commerce extensive power to investigate information related to "foreign adversaries" and transactions involving information and communications technology and services and to establish procedures for the investigation. The latter aimed to prohibit trading and using foreign information and communications technologies and services that would threat American national security, foreign policy and economy.

In November 2019, Russia's *Sovereign Internet Law* took effect, which authorized Russian government to pre-emptively disconnect servers in Russia from international Internet servers in state of emergency. *Sovereign Internet Law* substantively enhanced Russian government's administrative authority over cyberspace. When the United States and Russia tightened their grip on cyberspace, European countries deployed their overall cyberspace governance systems. For example, Italian Council of Ministers approved *National Cybersecurity Convenience Act*, ordering that domestic network information service providers must ensure the security of network, information systems and information technology service; otherwise, they would be liable to a heavy administrative fine.

### 8.4.3 Long Arm Jurisdiction is Frequently Used, Which Provokes International Concern

In cyberlaw, jurisdiction plays an important part in guaranteeing and regulating national cybersecurity. In cyber era, some countries employ long arm jurisdiction to deal with the supply chain of critical network infrastructure as well as the connection, contact and trade of Internet users, in order to enlarge the applicable scope of domestic laws. Based on technological advantages, several countries frequently use long arm jurisdiction and undermine national laws, hoping to replace international rules with domestic laws.

On May 15, 2020, the Bureau of Industry and Security of the U.S. Department of Commerce announced a major amendment to *Export Administration Regulation*, by which the United States further enlarged the scope of long arm jurisdiction with the help of "Direct Product Rule". By expanding the applicable circumstance of "Direct Product Rule", *Export Administration Regulation* further imposed strategic restrictions on enterprises in Special Entities List, in order to prevent them from obtaining strategically significant products made by specific American technologies outside the United States.[1] On May 22, 2020, the United States released *U.S. National Security, Export Controls and Huawei: The Strategic Context in Three Framings*. By revising "Direct Product Rule", the United States imposed licensing requirements on foreign-produced items when there was knowledge that they were destined for Huawei (or for affiliates appearing on the Entity List). As a result, foreign items, including chipsets that were produced anywhere from designs of Huawei using the U.S. Department of Commerce-controlled semiconductor manufacturing equipment

[1] https://www.commerce.gov/news/press-releases/2020/05/commerce-addresses-huaweis-efforts-undermine-entity-list-restricts.

or software, would be subject to the U.S. export licensing restrictions when there was knowledge they were destined for Huawei or one of its listed affiliates.[2]

## 8.5 New Technologies and New Business Forms Accelerate Iteration, the Risk-Prevention Mechanism Gradually Matures

As the Fourth Industrial Revolution advances, new technologies and new business forms constantly emerge and display a tendency of sustainable and rapid development. Traditional products and new technologies coalesce and boost industrial development and product iteration. The development of new technologies and new business forms widely meets the practical needs of the public, yet produces new risks. Many countries promote technological and industrial development via policy support and control the risk of technological applications via legislative means.

### *8.5.1 The Legislation of Digital Payment Incessantly Improves, the Legal Supervision of Blockchain Gradually Plays a Dominant Role*

In digital era, the digital trend of economic forms becomes evident, and digital payment models mature. To ensure the security and reliability of digital payment, many countries put the legislative work of digital payment on the agenda. Blockchain technology develops fast and turns into the central work of legal supervision in digital era.

In January 2020, Monetary Authority of Singapore (MAS) declared that *Payment Service Act* came to force. One of its aims was to heighten public confidence in electronic payment and to include new payment service like digital payment token service. In May 2020, Japan's *Payment Service Act* (PSA) and *Financial Instruments and Exchange Act* (FIEA) took effect. Among them, the amendment to *Payment Service Act* (PSA) raised relevant requirements for the application procedure, customer asset management, entrusted execution, securities margin trading, advertising and marketing of crypto asset exchange service providers, and specified activities and behaviors that were banned. The amendment to *Financial Instruments and Exchange Act* (FIEA) categorized crypto asset into "financial instrument". The investment rights and interests of partnership enterprises that were transferable via blockchain were viewed as security, which could be strictly regulated in line with Type-I security regulatory models. In this way, to provide Japanese investors with

---

[2] https://www.state.gov/wp-content/uploads/2020/05/T-Paper-Series-U.S.-National-Security-Export-Controls-and-Huawei.pdf.

this kind of crypto asset would be tightly investigated to avert regulatory defects. In the meantime, it was prohibited to use blockchain for deception and fraud, market manipulation and malpractice related to crypto asset.

In various countries, legal supervision and policy regulation place emphasis on blockchain technology. U.S. government attached much attention to the problems in the development and application of blockchain, with legislation done at federal and state levels. In July 2019, at the federal level, congressmen presented to U. S. Congress the legislative proposal *Blockchain Promotion Act* that proposed to organize a working group on blockchain at the federal level, clarify the technological definition of blockchain, promote the uniform criterion and give opportunities for use of blockchain at the federal level. State governments also set up corresponding legislative standards. For example, the State of Nevada signed four blockchain acts on June 7, 2019 and June 13, 2019 respectively: SB161, SB162, SB163 and SB164, which chose relatively lenient regulatory and legislative models to support innovation and investment. In January 2020, *Illinois State Blockchain Technology Act* entered into force, which acknowledged the legal effect of smart contracts and blockchain technology-based electronic signatures in business and sternly restricted the use of blockchain in copyable transaction records or copies of contracts.

### *8.5.2 Drawing the Blueprint of AI Development, Constructing the Moral Norms for AI*

AI displays great prospect and unpredictable risk in application. Various countries endeavor to fully take advantage of its positive effect and curb its negative effect. Besides, regulators hope to achieve the harmonious co-existence between AI and human.

1. Deploying AI Development Strategy

In order to gain an advantage and maintain their leading position, many countries foreground the strategic role of AI. In October 2019, Russia published *National Strategy for the Development of Artificial Intelligence for the period until 2030*, which confirmed the developmental goal of AI, clarified the principle in the development of AI application and incorporated the AI developmental strategy document into "Russian Federation Digital Economy" National Development Plan. On November 13, 2019, Singapore launched an 11-year National AI Strategy, which presented the vision, path and priority for AI future development in Singapore. The Strategy signified a great step in the vision of "Smart Singapore". On February 19, 2020, European Commission released *2020 Artificial Intelligence White Paper*, which aimed to improve the innovation in the field of AI in Europe and promote the development of ethical and trustworthy AI. Hoping to forge a "trustworthy AI framework", *2020 Artificial Intelligence White Paper* focused on three major goals: researching and developing human-centered technology, developing fair and competitive economy, and building open, democratic and sustainable society. In order to maintain their dominance in AI competition, many countries continually advance strategic layout in the AI field. As the systematic layout improves, relevant legislation is put on the agenda. This provides legal guarantee for the development of AI industries with specific rules.

2. Establishing AI Application Norms

In AI development, it is an important issue to regulate AI application, restrain AI application risks and promote AI friendly development in society. In November 2019, European Commission released a report entitled *Accountability for Artificial Intelligence and Other Emerging Technologies*, which analyzed the legal liability that arouse from AI. In January 2020, the White House issued *Guideline for Artificial Intelligence Application and Standards (*draft of memorandum). It deeply examined the industrial development of AI technology and the standards of AI application, and emphasized public trust & participation and truthful and scientific methods. As it claimed, comprehensive and effective risk assessment mechanisms needed to be established, taking into account the benefit & cost and flexibility of technological input, as well as the potential discrimination and inequality caused by technology. It valued information disclosure and transparency, as well as the security and coordination & cooperation among organizations in AI application. Due to the uncertainty of AI risks, various countries took a cautious attitude toward AI legislation and commonly regulated AI risks with guiding documents like guidelines. As AI technology matures, AI application risks emerge, and the legislative work that standardizes AI application keeps following up.

### 8.5.3 Internet of Things (IoT) Industry Opens up New Situation, Rules for Security Guarantee Initially Take Shape

The network stability and risk defense capability of IoT play a crucial role in ensuring security and application. In order to ensure the healthy development of IoT industry and protect the legitimate rights & interests of users, the security & application standards of IoT industry are started to be made, which mainly concern privacy security and cybersecurity.

1. IoT market keeps expanding and the protection of user information receives wide attention

As IoT continuously advances, the protection of personal data raises more requirements and seeks for institutional responses. According to the data released by nCipher Security, a data security solution provider, and Ponemon Institute, a survey institute, as digital programs like cloud data and IoT develop quickly and data volume and type multiply, the priority is given to the protection of customers' personal information.

On January 1, 2020, *California Consumer Privacy Act* (SB 327) and *Oregon State Internet of Things Devices Security Act* (HB 2395) took effect, which aimed to ensure IoT security and application and focused on the obligation of manufacturers. The two acts demanded that manufacturers undertook the responsibility of adding "reasonable security functions" to devices or physical objects that were connected directly or indirectly to the Internet. This intended to protect devices and the information they contain from unauthorized access, destruction, use, modification or disclosure.

2. IoT Risks gradually come out and security protection standards urgently need to be set up

IoT regulators and users are tremendously beset by the emergence and spread of IoT risks. Therefore, to set up standards on IoT application & security becomes an item on the agenda. On January 9, 2020, the U.S. Senate passed *Promoting Internet of Things Innovation and Development Act*, requiring U.S. Secretary of Commerce to convene a federal government working group that examined the current policies and judicial practice & IoT development, tracked IoT problems, revealed obstacles in IoT laws, and presented proposals and reports to U. S. Congress for better IoT connectivity. On 27 January 2020, the UK government announced that it would enact new laws to improve the security standard of IoT devices, in response to prominent IoT security. New laws would ensure that all IoT products sold in the UK met three security requirements:

(1) All IoT products must have a unique password, which could not be reset to any common factory default value.
(2) Manufacturers of IoT products must provide a public window for anyone to report security vulnerabilities.
(3) Manufacturers of IoT products must clearly inform of security updating schedule of IoT products.

3 Policy support for automatic driving is strengthened and technological risk prevention system is optimized

As automatic driving technology continuously develops, its implicit value in use is acknowledged, winning legislative and policy support in many countries. Automatic driving technology radically reforms traditional driving technology. The sustainable development of automatic driving industry plays a significant role in optimizing traffic structure and improving traffic efficiency. Many countries have realized the potential application of automatic driving and provided policy support. In January 2020, the United States officially released the guideline *Ensuring American Leadership in Automated Vehicle Technologies: Automated Vehicles 4.0*, which clarified the economic and social benefits of automatic driving and particularized various investment and promotion of the U.S. government as well as resources for automatic driving innovators. The guideline meant the United States gave great policy support to ensure its leadership in automated vehicle technologies. In order to promote the application of automatic driving technology, some countries commence the legislative work. In July 2020, German government launched landmark legislation. Kraftfahrt-Bundesamt (KBA) expressed that "New legal framework must establish a foundation for the current legislative work, allowing the standard operation of automated vehicles on public roads under special geographical conditions. Automated vehicles are widely applied to various use cases, which are unnecessarily specified. The flexibility requires us to take into account all forms of driving." Covering a complete legal framework, the draft was under revision and expected to be approved in 2021.

# Chapter 9
# International Cyberspace Governance

## 9.1 Outline

In 2020, the deep adjustment of relations among major countries produced superimposed effects with COVID-19, which intensified the uncertainty and vulnerability of international cyberspace governance. Relations among major countries deeply interfered with the process of international cyberspace governance. The tussle in cyberspace among countries submerged from the game of cybersecurity rule into the field of basic information technology, and extended to emerging technologies, with the geopolitical trend in cyberspace prominent. The relationship between non-state actors and state actors in international cyberspace governance became more complex, urging us to construct fair and just international cyberspace governance systems. Meanwhile, multi-party cooperation was promoted. In particular, since the outbreak of COVID-19, all parties came to realize that it was necessary to enhance their capacity of coping with risks and challenges via cooperation. They also boosted the construction of cyberspace order and the recovery of society and economy in the post-COVID-19 era.

In the face of complex and volatile international circumstances, international community strived to improve international cyberspace governance and forged consensus. The United Nations officially launched the "dual-track" negotiation process of Group of Governmental Experts and Open Working Group on Information Security to promote the discussion on international cyberspace rules. All parties continued to strengthen cyber content governance and facilitate information technology governance and standard setting. Particularly, the discussion on the security of ICT supply chains and the international standard of new technology deepened, and various countries took measures to bridge "digital divide", with the game of rules on digital economy intensifying. Major countries and regions in the world gave priority to cutting-edge technologies like 5G, AI and quantum computing in scientific & technological competition, attached great importance to cybersecurity, development and governance, and continually improved cyberspace policies and strategies. Developed countries and regions like the United States, the European Union, Japan, Germany



Chinese Academy of Cyberspace Studies, *World Internet Development Report 2020*,
https://doi.org/10.1007/978-981-16-9388-5_9

and the United Kingdom maintained their competitive advantages in cyberspace by means of technological development and rule-making. China actively promoted the construction of a fair and just international cyberspace governance system, and deepened and actualized the concept of community of shared future in cyberspace. Russia continually enhanced the construction of independent and controllable cybersecurity systems. Other developing countries and regions also voluntarily took advantage of the Internet to develop cutting-edge technologies and partook in the formulation of international rules in cyberspace.

## 9.2 Annual Highlights of International Cyberspace Governance

Information technology represented by the Internet develops vigorously, which profoundly transforms the landscape and form of international competition. In major countries and regions, the competition on cyberspace strategies and governance models increasingly intensifies. New technologies have not only become a key field in the competition among countries, but also created opportunities for transnational cooperation on cyberspace governance. COVID-19 aggravates prominent problems such as global backward governance systems, weak capacity and institutional inadequacy. The original system and mechanism of international cyberspace governance hardly accord with the status quo, and international cyberspace governance stays in a critical time.

### *9.2.1 The Outbreak of COVID-19 Exacerbates the Uncertainty and Vulnerability in International Cyberspace Governance*

As the biggest black swan event in the twenty-first century, COVID-19 not only intensified the existing conflict in international governance, but also posed new problems and challenges. In the outbreak of COVID-19, several countries launched "infodemic war", perfidiously abandoned multilateral systems, evaded international responsibilities and obligations, and seriously impaired global governance systems centering on international organizations. Cybersecurity incidents frequently emerged, and crimes that used the COVID-19 virus to launch malicious attack, Cyberextortion and identity thefts soared. Digital technology like big data, AI and cloud computing played a remarkable role in monitoring & analyzing the pandemic, tracing the origin of COVID-19, preventing & controlling COVID-19 and allocating resources. However, it was urgently needed to solve the protection of data privacy and the balance of public interests that came after, with the tension between the prominent demands of cyberspace governance and the absence of cyberspace rules taking turn for the worse. In a long term, many countries will concentrate on pandemic prevention & control and economic recovery. As a result, they probably attach less attention and invest less resources to international cyberspace governance, with the demands out of all proportion to the efforts they take. Acknowledging the uncertainty and vulnerability of international cyberspace governance, international community endeavors to promote cooperation and protect the peace, openness and development of cyberspace.

### 9.2.2 Geopolitics Enormously Affects the Direction of International Cyberspace Governance

Presently, major countries deeply re-shape their relations, and the world faces a transitional turn. The game between major countries in cyberspace is featured with composite means and diverse fields. In the application of 5G technology and Internet, there is a remarkable geopolitical trend. Globally, the ideology of de-globalization takes shape, and unilateralism and protectionism rage. Under such circumstances, more countries and regions accentuate technological sovereignty and digital sovereignty, deem cutting-edge technologies like AI, 5G, cloud computing and quantum communication key factors in national competitiveness, and take protectionism to raise self-reliance. Using the pretext of "national security" groundlessly, several countries launch critical technological "outage" and "blockade", wantonly ban foreign Internet apps, forge exclusive technological standards & international rule systems, coerce other countries into "standing in line" and "choosing camp", and destroy the security and order of ICT supply chains. As a consequence, there are more risks for cyberspace to split into multiple criteria and systems. Cybersecurity turns into an important factor to consider in national security and geopolitics. Individual countries take the means of "netsplit" to defend national security. APT attack coincides with geopolitical tussle. With cyberwar as an open military approach, the militarization of cyberspace intensifies, and the strategic stability of cyberspace faces a great challenge. In spite of the fact that geopolitics prominently influences various issues on cyberspace governance or impedes the progress of cyberspace governance, all parties of international community seek to hedge against negative effects, enlarge consensus and maintain the development and stability of cyberspace.

### 9.2.3 The Model of International Cyberspace Governance Needs to Be Adjusted and Reformed

Currently, world multi-polarization, economic globalization, social informatization and cultural diversity develop more deeply, international power landscape evolves more quickly, and various actors enhance their subject roles. COVID-19 objectively spurs the return of "statism", and state actors play a more significant role. How to counterbalance the responsibility of global governance with national security & interests means an enormous challenge to state actors. Information revolution accelerates the spread of power and re-forges the influence of governing bodies. Non-state actors, e.g. super IT corporations, impact more deeply on world politics and international security. They play an more active role in the governance of new technologies & applications and the formulation of rules for behaviors in cyberspace, becoming an important force in promoting the openness and cooperation in cyberspace. Traditional international organizations attempt to take part in the formulation of rules on digital economy and cyberspace behaviors, conditioned by backward governance structure, geopolitics and de-globalization. Geopolitics and technological development enormously affect the functus officio of Internet communities. The capability of technological communities in promoting innovative development and technological protocol is partially constrained by internal policies and regulations,[1] with multi-stakeholder governance models challenged. Under the current circumstances, various actors dish out advice on the establishment of international rules and governance systems in cyberspace. For instance, China proposes the concept of community of shared future in cyberspace, which effectively explores how to advance the establishment of a fair and just international cyberspace governance system, with wide support from international community.

## 9.3 New Progress Made in Issues Related to International Cyberspace Governance

In 2020, all parties still discussed and formulated rules on cybersecurity, digital economy and AI via multilateral platforms, making rapid progress in bilateral and regional rules. The steady development of technologies like 5G, AI, big data and quantum computing diversified the topics on cyberspace governance and created more space for cooperation among all parties.

### *9.3.1 Rules for International Cyberspace Governance Continue to Advance*

The United Nations plays a more important role in formulating international rules in cyberspace. In September 2019, the United Nations firstly launched the "dual-track" negotiation process of Group of Governmental Experts and Open Working Group on Information Security. In December 2019, General Assembly of the United Nations passed Against the Criminal Use of ICTs and officially launched the negotiation of global convention on cybercrimes. All parties widely applauded the application of

[1] Cai Cuihong. "Cyberspace Governance in the Era of Global Change". Exploration and Free Views, 2019 (1).

international laws into cyberspace and required the respect for Charter of the United Nations and cyberspace governance within the UN framework. However, it was necessary to enlarge the consensus among international community on the application of international laws and Charter of the United Nations to cyberspace, the scope of sovereignty in cyberspace and the means to safeguard it, rule on emerging technology, and the origin-tracing of and response to cyberattack. France, Estonia and the Netherlands made clear their positions and presented their proposals on the application of international laws. For example, Estonia emphasized the application of international laws to cyberspace, appealing to all countries to cautiously use cyber-weapons to safeguard cyber sovereignty. The Netherlands stressed that openness, freedom and integrity of cyberspace ought to be ensured within the framework of the United Nations.[2]

Other stakeholders in cyberspace also made great efforts to establish international rules on cyberspace. In November 2019, World Wide Web Foundation, founded by Tim Berners-Lee, father of World Wide Web, released Contract for the Web, which proposed "Nine Basic Principles" on Internet protection at governmental, corporate and individual levels, in terms of online fake news, infringement on privacy, online violence and political manipulation, in order to promote the development of Internet.[3] At Paris Peace Forum held in November 2019, Global Commission on the Stability of Cyberspace published Advancing Cyberstability. With "Eight Norms" that maintained cyberspace stability, it established common norms for both state and non-state actors in cyberspace.

### *9.3.2 The Game of Making Rules for Digital Economy Intensifies*

Now, the game that centers on digital tax, cross-border flow of data and rule on digital trade becomes fiercer. Major countries and regions in the world hope to forge their voice and influence concerning digital rules by making regional and bilateral & multilateral agreements.

(1) The collection of digital tax accelerates, and digital divergence between the Americas and Europe appears. By June 2020, 22 countries and regions had implemented or proposed digital taxes in various forms.[4] G20 and OECD served as two major international platforms for the discussion of rules for the collection of digital tax. OECD was committed to coordinating the negotiation among 137 countries and regions in the world and to promoting the

[2] Wang Zheng. "New Situation in the Formulation of Global Cyberspace Rules under UN 'Dual Track System'". China Information Security, 2020 (1).

[3] Guo Feng and Huang Xiaoyi. "Contract for the Web & The Construction of International Cyberspace Rules". China Information Security, 2020 (1).

[4] "The Dispute over Digital Service Tax in the United States and Europe Escalates". See http://paper.people.com.cn/rmrbwap/html/2020-07/20/nw.D110000renmrb_20200720_7.

revision of the existing cross-border tax rules, in order to address tax-system challenges that arose from the development of digital economy and support the goal that G20 Finance Ministers Meeting formulated unanimous standards on the collection of digital tax by the end of 2020. In terms of revenue estimation and taxation model, there was a deadlock between the United States and other countries, because the United States reckoned that digital tax deliberately targeted American companies[5] and violated international trade rules, announcing "The Investigation under Section 301 of The Trade Act of 1974", against 10 countries and regions including the European Union and the United Kingdom and withdrawing from OECD digital tax negotiation. Besides, in July 2020, the European Court of Justice overruled Europe-US Privacy Shield Data Transfer Agreement, compelling them to re-negotiate privacy shield transmission framework.

(2) Digital trade rule-making displays a regional and fragmentary trend. No substantive progress was made in multilateral negotiation, yet that was made in negotiation over digital trade rules at bilateral and regional levels. For example, in US-Mexico-Canada Agreement (USMCA) signed by the United States, Canada and Mexico, a section concerned digital trade. The United States and Japan signed US-Japan Digital Trade Agreement.

### *9.3.3 Consensus is Gradually Formed in the Governance of Cyber Content*

In recent years, international community attached increasing importance to the governance of cyber content. Since March 2020, China has formally implemented Regulations on Ecological Governance of Cyber Information Content, which highlighted the subjective initiative of multiple subjects of "government, enterprise, society and netizen" who participated in the ecological governance of cyber, and mainly standardized the rights and obligations of cyber information content producers, cyber information content service platforms, cyber information content service users and cyber industry organizations in the ecological governance of cyber. In May 2020, France passed a law that prohibited online hate speech, requiring platforms like Google, Twitter and Facebook to delete tagged hate content within 24 h and terrorist propaganda within one hour. Otherwise, they would face a fine of up to 1.25 million euros. In October 2019, Singapore's Protection from Online Falsehoods and Manipulation Act took effect. The Act gave Singaporean government the right to order individuals or online platforms to correct or delete fake news with negative impacts on public interest. Offenders would be sentenced to up to 10 years in jail and fined up to 100,000 Singapore dollars. In August 2019, members of Group of Seven, excluding the United

[5] "The U.S. is hurtling toward another trade war—but this time it isn't with China", 06/18/2020, https://www.politico.com/news/2020/06/18/europe-us-digital-tax-trade-war-328338.

States, signed Charter for a Free, Open, and Safe Internet,[6] which aimed to crack down on illegal and harmful content on online platforms. In the meantime, large Internet corporations such as Facebook, YouTube, Twitter and TikTok strengthened their platform governance and tightened the combat against online false information.

### 9.3.4 The Governance and Standard-Making of Information Technology Keeps Improving

On the one hand, Internet Corporation for Assigned Names and Numbers (ICANN) improves its accountability and transparency, organizes the research and formulation of solutions to advance root server system governance models, enlarges multiple participative parties and clarifies entry and exit mechanisms of operating institutions. ICANN also promotes privacy protection compliance and establishes working groups to quicken the docking with the European Union in privacy protection rules on New-Generation WHOIS Lookup System. On the other hand, international community tightens technological governance on 5G and AI with respect to technological standards and application norms. Telecommunications corporations, International Telecommunication Union and International Organization for Standardization of Mobile Communications continue to advance the formulation of standards on 5G technology. In the field of AI norms, in June 2019, China National New Generation Artificial Intelligence Governance Committee released New Generation Artificial Intelligence Governance Principles, which raised eight principles (i.e. harmony & friendliness, fairness & justice, inclusiveness & sharing, respect of privacy, security & controllability, co-undertaking of responsibility, openness & collaboration and agile management) and required to develop responsible AI. In October 2019, the U.S. DoD Defense Innovation Board published AI Principles: Recommendations on the Ethical Use of Artificial Intelligence, which proposed five principles, i.e. accountability, fairness, traceability, reliability and controllability. In February 2020, the European Union released White Paper: On Artificial Intelligence—A European Approach to Excellence and Trust, which aimed to ensure that AI application accorded with EU values, basic rights and ethical principles. In general, AI rule stays at early stage of exploration rather than substantive stage of legislation.

### 9.3.5 Talks on the Security of ICT Supply Chain Deepen

In recent years, as the competition on the application of emerging technology and the development of digital economy became increasingly fierce, various countries paid the most attention to the security of ICT supply chains, hoping to strengthen

[6] Charter for a Free, Open, and Safe Internet, on 21 August 2019. https://www.entreprises.gouv.fr/files/files/directions_services/numerique/Charter-for- a- free-open-and-safe-Internet.pdf.

the governance of the security of ICT supply chains. A variety of actors, including governments, transnational institutions and industrial & academic circles, actively contributed ideas to the governance of the security of ICT supply chains.

In the negotiation of the UN Open Working Group, Chinese government clearly stated its views on the security of ICT supply chains.

(1) No country shall use its own advantage to impair the security of ICT & service supply chains in another country.
(2) All countries should forbid suppliers of IT products and services to set up back doors when they provide products, by which they illegally obtain user data and control & manipulate user systems and equipment. No suppliers shall buck for illegitimate benefits or compel users to update or upgrade operating systems. Suppliers shall undertake to promptly notify their partners and users of any serious security loophole or defect they have discovered.
(3) All countries shall maintain a fair, just and non-discriminatory business environment and shall not restrict normal ICT development & cooperation, market entry of ICT products and export of high-tech products with the abuse of "national security".[7]

India proposed the standard of "free of supply chain tension". Iran also mentioned that "all countries enjoyed equal rights of supply chains, including the research & development, manufacturing, use and transfer of TCT products and services", arguing that the equal participation in supply chains meant no country could exclude another from supply chains.[8] International think tanks also sped up the research on the norm of ICT supply chain security. The East–West Institute, for example, advised to ensure ICT supply chain security and counteract ICT nationalism by improving transparency and accountability.[9] Carnegie Endowment for International Peace suggested that companies and governments ought to improve the reliability and integrity of ICT supply chains according to trust, responsibility, transparency and understanding. In its view, relevant rules were made based on informal international initiatives or international documents and embodied by documents on bilateral or multilateral mechanisms.[10] Presently, various actors have wide divergences on their responsibility and obligation in the security of ICT supply chains, which are politicized by some countries. Consequently, it is hard to reach a consensus in a short time.

[7] OEWG Chinese Position File, https://www.un.org/disarmament/wp-content/uploads/2019/09/china-submissions- oewg-ch.pdf.

[8] Chen Hui and Xia Yuqing. "Security Situations and Countermeasures of International Supply Chains in the First Half of 2020". China Information Security, 2020 (7).

[9] EastWest Institute, "Weathering TechNationalism: A Security and Trustworthiness Framework to Manage Cyber Supply Chain Risk", https://www.eastwest.ngo/sites/default/files/ideas-files/weathering-technationalism.pdf.

[10] ARIEL (ELI) LEVITE, ICT Supply Chain Integrity: Principles for Governmental and Corporate Policies, https://carnegieendowment.org/2019/10/04/ict-supply-chain-integrity-principles-for-governmental-and-corporate-policies-pub-79974.

### *9.3.6 International Community Strives to Bridge the Digital Divide*

As the access and usage of global mobile and fixed broadband network continue to rise, various countries and regions unevenly develop in Internet popularization, infrastructure construction, technological innovation & creation, security-risk prevention and digital-skill mastery. Moreover, COVID-19 highlights the importance of digital technology and urgently requires to bridge the digital divide via science, technology and innovation. At the United Nations level, International Telecommunication Union has been supporting many countries in the improvement of policies & standards, promoting the popularization of telecommunications infrastructure and broadband access, and playing an important role in bridging the digital divide. In April 2020, the United Nations released the research report The COVID-19 Crisis: Accentuating the Need to Bridge Digital Divides, which stated that a multilateral coordination mechanism was needed to address digital challenge (e.g. making new policies and regulations), to bridge the existing and enlarging digital divide and to enable more countries to benefit from digital dividend. In the outbreak of COVID-19, the United Nations launched a technology partnership initiative to help developing countries ameliorate their capacity in allocating resources, expanding production and obtaining critical health equipment by using digital technologies. In June 2020, António Guterres, Secretary-General of the United Nations, officially proposed "Roadmap for Digital Cooperation", with seven tasks, i.e. promoting digital general connection, advancing digital technology as public products, ensuring the inclusiveness of digital technology for all the people, supporting the construction of digital capacity, protecting human rights in digital fields, addressing the challenge of AI and forging digital trust and security. Besides, António Guterres was considering appointing a technology envoy to foster the establishment of the UN Framework for Global Technological Cooperation. At the national level, China promoted the construction of "Digital Silk Road" based on the principle of consultation, contribution and shared benefits, which played a vital role in promoting the construction of digital infrastructure in countries and regions along Central Asia, Africa and Latin America as well as bridging the digital divide in the world.

## 9.4 Cyberspace Governance in Some Typical Countries and Regions

In 2020, in the face of complex and unstable international landscape, some countries and regions attached more importance to cyberspace development & security, improved cyberspace rules & legal systems according to their reality, increased the investment in cutting-edge technologies like AI, 5G and quantum computing, strengthened the protection of personal information, accelerated the construction of cybersecurity guarantee systems and enhanced voice and influence in cyberspace.

Noticeably, in order to maintain their dominant position, some countries generalized issues on cybersecurity, which generated negative spillover effects and cast a shadow on the openness & development of cyberspace.

### *9.4.1 The United States*

In 2020, the United States continually improved policy & law and institutionalization on cyberspace to maintain its dominance in cyberspace.

(1) Improving the establishment of organizations and strengthening deterrence in cyberspace. U.S. Cyberspace Solarium Commission released Layered Cyber Deterrence Strategy, which strengthened the ideas of "continuous engagement" and "defensive prepositioning", aiming to enhance U.S. aggression and deterrence in cyberspace. To improve the establishment of cybersecurity organizations and clarify the responsibility of cybersecurity departments, U.S. National Security Agency founded Cybersecurity Directorate. Optimizing ODNI's (Office of the Director of National Intelligence) function on cybersecurity, the United States set up IC Cyber Executive, which comprehensively strengthened the capabilities of intelligence collection, cyber defense and cyber warfare. Simultaneously, the United States quickened Indo-Pacific military-cyber deployment and launched cyber offensive & defensive drill.

(2) Enhancing support to ensure U.S. leading role in cutting-edge technology. The U.S. government regards 5G, AI and quantum computing as key new technologies to win in the future. The U.S. House of Representatives and the U.S. Senate intensively approved relevant acts, like National Strategy To Secure 5G of the United States of America, The Secure 5G and Beyond Act of 2020 and Secure and Trusted Communications Networks Act to secure U.S. competitive advantages and security in 5G. The U.S. government released American Artificial Intelligence Initiative: Year One Annual Report, Artificial Intelligence & Quantum Information Science R & D Summary: Fiscal Years 2020–2021 and A Strategic Vision for American's Quantum Networks, which substantially increased the investment in AI and quantum computing. The United States also intensified the rule-making on emerging technologies. U.S. Office of Management and Budget published Guidance for Regulation of Artificial Intelligence Applications, with a large number of advice on regulating AI norms for federal agencies and industrial sectors. Besides, the United States joined G7 Global Partnership on AI and played an active role in making AI international rules.

(3) Emphasizing "America First" and maintaining U.S. dominance in cyberspace. In 2020, the United States generalized the idea of national security, pursued unilateralism and trade protectionism, and employed "long arm jurisdiction", "Entity List" and "Clean Network Program", which politicized digital technology. For example, in the development, application and cooperation of 5G, the United States took discriminatory stance and squeezed competitors, which

not only breached the code of fair competition, but also inhibited the inclusiveness of new technologies for human, with the openness and cooperation in cyberspace stunted.

### 9.4.2 China

China accelerates the construction of comprehensive cyberspace governance systems, diversifies ideas and views on international cyberspace governance, launches international exchange and cooperation and establishes partnership in cyberspace. In October 2019, concept paper on Jointly Build a Community with a Shared Future in Cyberspace and result document Wuzhen Outlook 2019 were published in the Sixth World Internet Conference, and result document Sovereignty in Cyberspace: Theory and Practice was released in the forum of the Sixth World Internet Conference. These further expounded China's ideas and views on international cyberspace governance.

(1) Promoting international cooperation in cyberspace. China actively partakes in UN Open-ended Working Group and Group of Governmental Experts Course on information security with constructive stance, boosts the discussion on UN Convention on Combating Global Cybercrime, voluntarily hosts open forum of UN Internet Governance Forum, and participates in the work of the multistakeholder advisory group of the open forum. China deepens the cooperation in cyberspace with countries and regions like Russia, the European Union, the United Kingdom, France, Germany, Japan, South Korea and Association of Southeast Asian Nations, with dialogues and exchanges on cutting-edge technology, digital economy, cybercrime, communications-infrastructure construction and counter-terrorism in cyberspace. China deeply engages in cybersecurity course under the frameworks of Shanghai Cooperation Organization and ASEAN Regional Forum, as well as the discussion on digital and cyber issues within G20 and Asia–Pacific Economic Cooperation. China launches Global Initiative on Data Security, appealing to all countries to take precautionary measures to curb the damage to or theft of major data on critical infrastructure and the infringement on personal information by using information technology, to combat large-scale monitoring of other countries with the abuse of information technology, and to allow domestic enterprises to store data overseas. Besides, enterprises shall not set up back doors in their products and services. Committed to maintaining global data security, the Initiative provides a blueprint for making relevant global rules.

(2) Accelerating the construction of comprehensive cyberspace governance systems. In line with international consensus and trend, China improves policies and rules on content governance, personal information protection, Internet finance and cybersecurity and releases laws and regulations like Regulations on Ecological Governance of Cyber Information Content and Cybersecurity

Review Measures, in order to advance the formulation of laws on personal information protection and data security.

(3) Fostering international cooperation in the fight against COVID-19 with digital technology. After the outbreak of COVID-19, China actively actualized the consensus reached at the G20 Summit and conducted online exchanges on the prevention & control of COVID-19 with South Korea, Japan, Russia, the United States, Germany and South Africa as well as several international and regional organizations. Chinese Internet companies also provided intelligent diagnostic technology to medical institutions in Japan and Italy and remote video conferencing systems to Cambodia, so as to jointly address the challenge of COVID-19.

### *9.4.3 Russia*

Russia makes substantive progress in constructing autonomous information security systems and takes efforts to build an autonomous and controllable cyberspace ecosystem.

(1) Improving the autonomous and controllable capability in cyberspace. Russia successfully carried out the test of "Internet blackout" and set up Russia General Communications Network Supervision Center, to ensure Russian government's monitoring of Internet information exchange routes and the smooth operation of Russian Internet in the case of sustainable threats.

(2) Determining the principles of personal-data localization. In December 2019, Russian President Vladimir Putin signed Federal Law No. 405, which further enhanced the penalty on illegal acts in the field of personal data and information dissemination and raised the maximum fine for the breach of data localization to 18 million roubles.

(3) Actively participating in the course of international cyberspace governance within the UN framework. Russia positively urged the United Nations to launch the negotiation on global convention on cybercrimes. Russia also suggested that presently various countries made joint efforts to formulate responsible national code conduct in cyberspace, fully considering the advice from some countries on strengthening national sovereignty in cyberspace, applying ICT exclusively for peaceful purpose, ensuring the integrity of supply chains and preventing the militarization of cyberspace.

### *9.4.4 The European Union*

The European Union strives to strengthen technological sovereignty, promote European digital transformation and enhance EU's influence in digital governance.

(1) Strengthening the top-level design and improving the digital strategic autonomy of Europe. In February 2020, European Commission released an overall planning on Europe's approach to digital era as well as two documents White Paper on Artificial Intelligence and European Data Strategy, prioritizing the task of building "digital future of Europe", promoting EU's innovation on new technology and new application like AI, 5G, quantum computing and EU's power in rule-making, and enhancing Europe's competitiveness and strategic autonomy in digital transformation.

(2) Emphasizing 5G security and optimizing cybersecurity governance framework. According to the reality in Europe, the European Union successively published EU Coordinated Risk Assessment of the Cybersecurity of 5G Networks and EU Toolbox for 5G Security 2020, planning to make security standards for 5G networks. In March 2020, six countries, i.e. Estonia, Poland, Croatia, the Netherlands, Romania and Lithuania, established European Union Quick Reaction Cyber Force to improve their response to cyberattack. In August 2020, European Commission issued EU Security Alliance Strategy 2020–2025, which gave priority to maintaining future-oriented security environment, tackling growing threat, protecting European citizens from terrorism and organized crimes, and forging a strong European security ecosystem, with plans raised in detail.

(3) Tightening digital governance and enlarging the influence of digital rules. The European Union continually improved General Data Protection Regulation (as a guideline) and strengthened data protection and privacy regulation. It released Code of Practice on Disinformation, which required tech giants to submit monthly disinformation reports. It steadily advanced the formulation of Digital Services Act to clarify the responsibilities and obligations in digital service. In addition, the European Union nominally supported that the United Nations should play an more important role in international cyberspace governance, especially in the establishment of cybersecurity rules, stating that the existing international law, including international humanitarian law, was applicable to cyberspace.

### *9.4.5 Japan*

Japan quickens the development of cutting-edge technology and strengthens the protection of data, in an attempt to set up a bellwether for digital trade rules.

(1) Amplifying the investment, planning and governance in cutting-edge technology. In June 2019, Japan released 2019 Artificial Intelligence Strategy, which set tasks and goals for Japan's AI development, boosted the enactment of international standards and the training of talent, and guided AI technology R&D and industrial development. In January 2020, Japan released 6G Technology Overall Strategy (Draft), expressing that Japan would promote the research and development of 6G technology by means of financial support

and tax incentive. In five years, Japan strives to achieve a critical technological breakthrough and plans to realize the practical use of 6G communications in 2030. In March 2020, Japanese financial regulator proclaimed the start of Global Blockchain Governance Initiative Network to promote global cooperation in blockchain regulation.

(2) Perfecting the rules on personal data protection and strengthening the governance of tech giants. In November 2019, Japan passed Digital and Platform Transaction Transparency Act, which required Internet corporate platforms to disclose the rules for setting search results and improve the disclosure transparency of their transaction information. In June 2020, Diet of Japan passed an amendment to Personal Information Protection Law, which required platforms to obtain users' consent prior to providing personal data like Internet browsing history to a third party, in order to further strengthen the protection of public personal information.

(3) Setting up a bellwether for international rules on digital trade. In October 2019, Japan and the United States signed U.S.-Japan Digital Trade Agreement. As consented, Japan and the United States would cancel tariffs on electronic products like videos, music and e-books, oppose data localization, resist the mandatory disclosure of source codes and algorithms, and guarantee the barrier-free transmission of data between them, hoping to improve the level of U.S.-Japan digital trade.

### *9.4.6 The United Kingdom*

Under the guidance of "Global UK" and "Digital UK" strategies, the United Kingdom spurs economic and social development, improves cybersecurity and enhances global influence in cyberspace via digital innovation.

(1) Initiating "Digital Market" strategy. The strategy proposes to censor major online advertising platforms like Google and Facebook, focusing on the source where their market power comes, the way they collect and use personal data, and whether the competition on digital advertising benefits consumers or not.

(2) Promoting international digital cooperation. In October 2019, the United Kingdom and the United States signed bilateral cooperation agreement Clarifying Lawful Overseas Use of Data Act, which allowed law enforcement departments to request data service directly from tech companies in the UK and the US to removes legal barriers to the timely collection of electronic evidence in the UK and the US. In June 2020, the United Kingdom released new future technologies & trade strategies, aiming to foster cross-border data flow between the UK and the Asia–Pacific Region.

(3) Strengthening cybersecurity defense capability. The United Kingdom establishes national cyberforce to crack down on terrorist groups, hostile countries and organised criminal groups that threaten the UK. In the outbreak of

COVID-19, UK National Cyber Security Centre launched cybersecurity initiative, which provided practical advice and guidance to protect devices, accounts and passwords from cyberattack. The UK government also called for more investment in building cybersecurity capability and respect for the application of international laws in cyberspace.

### *9.4.7 France*

France endeavors to enhance its influence and voice in international cyberspace governance and realize Europe's strategic autonomy in cyberspace. In international governance, in September 2019, Ministère des Armées released White Paper on Declaration of International Law in Cyberspace: Law on Cyber Operations in Peacetime, which elaborated on French stance on the use cases of Charter of the United Nations and International Humanitarian Law in cyberspace in times of peace and armed conflict. As the White Paper states, France reserves the right to beat back any cyberattacks that probably threaten international laws and the right to use self-defence against cyberattacks that probably evolve into armed aggression. In digital currency, France views that currency in private hands will endanger national monetary sovereignty and breed problems like money laundering, terrorism, financing and market dominance. France proposes to establish European public digital currency, signaling the initial stage of digital currency in France. Besides, France plays an active role in advocating and promoting the collection of global digital tax, reaffirming that France would officially start to levy digital tax at the end of 2020, irrespective of whether international agreements on digital tax collection would take shape. Meanwhile, France and Germany jointly gave support to the establishment of the "Gaia-X" cloud computing system to reduce the reliance on the United States and realize the autonomy and reliability of European data infrastructure.

### *9.4.8 Germany*

Germany encourages digital sovereignty, strengthens the construction and support on emerging technology, and enhances the guarantee in cybersecurity.

(1) Enhancing the guarantee in cybersecurity. Germany actively participates in cyberwar drill of North Atlantic Treaty Organization to improve the capability in cyberwar. Germany plans to revise IT-Sicherheitsgesetz to improve hardware security and to establish ATHENE, the largest national applied cybersecurity research center in Europe, to enhance the guarantee in cybersecurity.
(2) Increasing the investment and support on cutting-edge technology. In September 2019, Germany allocated 500 million euros for AI research and application. It also planned to provide a fund of 128 million euros to German AI research institutions by 2022. Additionally, Germany released

blockchain strategy, hoping to use blockchain technology to promote the digital transformation of German economy and society.

(3) Raising the level of cyberspace governance. At the 14th UN Internet Governance Forum, Angela Merkel, German Chancellor, emphasized multilateralism and digital sovereignty in cyberspace, appealing to reconstruct cyber order and establish interconnected cyberspace under the UN framework. Germany also tightened the regulation of giant tech corporations. In January 2020, German Federal Cartel Office required to launch new mechanisms to effectively restrict large tech corporations and protect the market position of small enterprises and the rights & interests of consumers.

### *9.4.9 India*

India takes rigid measures on data localization to strengthen cyberspace governance. In December 2019, Indian cabinet passed Personal Data Protection Act, which ruled that companies must store sensitive personal data on servers in India and determined the principle of data localization in legislative form. Indian government frequently launched "Internet blackout" to quell domestic conflicts and riots. In December 2019, Indian government decided to abrogate Article 370 of Indian Constitution and release The Citizenship Amendment Act 2019 in Kashmir, which triggered in violent protest in some regions. Indian government ordered to suppress the riot by "Internet blackout". India plans to improve its competitiveness in cyberspace by altering regulatory rules. In January 2020, Indian government proposed to revise Information Technology Intermediaries Guidelines (Amendment) Rules 2018 (draft) to strengthen the content supervision on large social media corporations. In June 2020, India banned the use of Chinese apps, on the pretext of "undermining national security and public order". In August 2020, Indian government backed relevant acts that shared non-personal data in an obligatory way. India takes part in international cyberspace governance, arguing that the use of international law in cyberspace must be discussed in the context of international security, to ensure that all countries equally discuss legal issues and matters organized by the United Nations. India proposes to establish a permanent body within the United Nations or a subsidiary body of UN Open Working Group, in order to formulate the rule, standard and principle of responsible conduct in cyberspace and to determine the implementation and application of international laws.

### *9.4.10 Australia*

Australia stresses personal data protection and cybersecurity, promotes the governance of new technologies like blockchain, and actively deepens its relations with the United States in cyberspace. Australia successively approved Competition and

Consumer Data Right Rules and Compliance and Enforcement Policy for the Consumer Data Right, which determined and clarified issues like right to data portability and consumer data security, as well as the construction of data sharing framework under the open banking model. In August 2020, Australian government issued Cybersecurity Strategy, which planned to invest $1.67 billion in ten years to improve the capability in cybersecurity guarantee and law enforcement, with a vision for governments, enterprises and communities. In other words, governments should strengthen the protection of individuals, businesses and key infrastructure; enterprises should guarantee the security of products and services; and communities should raise public awareness of cybersecurity. In March 2020, Australian government published National Blockchain Roadmap: Progressing towards a Blockchain-Empowered Future, attempting to strengthen the regulation, skill training and capacity building in blockchain industry and enhance the competitiveness of blockchain industry. To deepen the collaboration with the United States in cyberspace, Australia intends to amend Telecommunications Interception and Access Act, strengthen the cooperation with the United States in cyberlaw enforcement, and side with the United States in the development of 5G and the course of international rules in cyberspace.

### *9.4.11 Latin America*

In Latin America, mobile Internet develops quickly. Various countries take efforts to develop cutting-edge technologies like AI and IoT. Latin America is beset by weak infrastructure in cybersecurity and uneven regional development. Brazil plays an active role in international cyberspace governance. In June 2019, Agência Nacional de Telecomunicações (Anatel) approved "Telecommunications Network Structure Plan" to enlarge broadband access. In July 2019, Brazilian Senate sanctioned the inclusion of data protection in digital platforms in the list of basic rights and individual & citizen guarantee ruled in Brazilian constitution, and established Digital Governance and Information Security Board. In infrastructure management, Brazil released national IoT programme and established an advisory body called IoT Office, to promote IoT development and implementation in all sectors in Brazil. Argentina takes cybersecurity seriously, improves the level of data governance, publishes Cybersecurity Strategy, and establishes Cybersecurity Committee. This clarifies strategic objectives of cybersecurity in the framework of cybersecurity supervision, the protection of critical information infrastructure and the reinforcement of international cooperation. Argentinean Digital Protection Agency published several resolutions to specify the operative norms on personal data protection. Chile also actively promotes law-making on data protection, and deepens the dialogue and cooperation with the European Union on data protection and cross-border flow. Besides, Chile joins Convention for the Protection of Individuals with regard to Automatic Processing of Personal Data of European Council as an observer.

### 9.4.12 Africa

In Africa, the Internet starts late yet disseminates fast. South Africa and Nigeria take the lead in cyber governance in Africa. In July 2019, South African government issued Draft of Cybercrimes Regulation, which ruled stealing and obstructing data as a crime. South African government also tightens its grip on the regulation of cryptocurrency. In April 2020, South Africa's largest financial regulators jointly released policy paper, giving suggestions on cryptocurrency regulation. Nigeria attaches great attention to data protection. After Nigeria Data Protection Regulation 2019, Nigeria released The Draft of Framework for the Implementation of Data Protection, which further specified data protection regulation. The African Union accelerates the course of cyberspace governance. In October 2019, a ministerial meeting on communications information was held in the African Union, which acknowledged the role that digital technology and innovation played in achieving the goals of 2063 Agenda and the United Nations Sustainable Development. The meeting proposed to forge African common ground, coordinate and set the roles of all stakeholders, and share most fruitful experiences and practices, by using PRIDA (Policy and Regulation Initiative for Digital Africa) digital platform.

The world today is marked by changes unseen in a century. New-round scientific & technological revolutions and industrial reforms pose unprecedented opportunities and challenges to the international community. With the development of new technologies, applications, business forms and models, national competition is catalyzed or intensified. The openness of cyberspace is enormously impacted and international Internet governance systems are at a critical juncture. Under such circumstances, international community ought to put aside minor differences so as to seek common ground on the basis of mutual respect, communication and collaboration, in order to pursue common welfare, address common challenges, promote technological progress and achieve effective governance. In this way, the Internet will better benefit people of all countries in jointly building a community of shared future in cyberspace.

# Postscript

While compiling World Internet Development Report 2020, we are deeply impressed that today, as IT revolution keeps evolving, global IT innovation remains amazingly active. The Internet fundamentally transforms human's way of production and life. Particularly, as the COVID-19 rages globally, the Internet proves increasingly important, with the world today as a community of shared future, where humans are interdependent. The joint contribution and shared benefits in the global Internet mean to us a historical opportunity and historical mission. In World Internet Development Report 2020 (hereinafter referred to as "The Report"), we aim to comprehensively represent the status quo of world Internet development in 2020, explicate the global trend of Internet development from Chinese perspective, and better achieve the goals, i.e. joint efforts made to advance cyberspace development, protect cybersecurity and participate in cyberspace governance and share benefits.

During the compilation of The Report, we receive guidance and support from Office of the Central Cyberspace Affairs Commission (hereinafter referred to as "OCCAC"). Leaders of OCCAC give us specific guidance on The Report. Relevant ministries & commissions, bureaus & units of OCCAC as well as Office of the Provincial Cyberspace Affairs Commission provide us with strong support for necessary data and materials. The Report is launched by Chinese Academy of Cyberspace Studies (CACS), and co-organized & co-compiled by National Computer Network and Information Security Administrative Center, Beijing University of Posts and Telecommunications, National Industrial Information Security Development Research Center, Peking University, Tsinghua University, China Academy of Information and Communications Technology, and Beihang University. Experts and scholars, including Fang Binxing, Zhou Hongren, Wang Yimin, Shi Anbin, Du Yuejin, Chen Kai and Zhang Li, contribute their valuable suggestions. Main contributors are: Xia Xueping, Fang Xinxin, Li Yuxiao, Xuan Xingzhang, Zou Xiaoxiang, Liao Jin, Cheng Yifeng, Jiang Wei, Jiang Yang, Nan Ting, Wang Zhongru, Wang Hailong, Yuan Xin, Li Xiaojiao, Jia Shuowei, Chen Jing, Xu Yu, Xiao Zheng, Wu Wei, Zhao Gaohua, Zhao Lei, Cai Yang, Sun Luman, Li Wei, Wang Hualei, Wang Liying, Chen Fengxian, Huang Peng, Mou Chunbo, Sun Ying, Jin Zhong, Chong Dandan, Wu Yanjun, Zhang Xuejun, Wen Jingwei, Shi Xiaolong, Chen Xiaogang,



Chinese Academy of Cyberspace Studies, *World Internet Development Report 2020*,
https://doi.org/10.1007/978-981-16-9388-5

Gao Taishan, Wang Wei, Zhang Nan, Huang Meiyin, Wang Peinan, Xu Yuan, Lou Shuyi, Wang Shiwen, Wen Baihua, Guo Feng, Zhao Jingwu, Zhou Ruijue, Mierbeke Sailek, Li Yan and Dai Lina.

The Report, albeit successfully published thanks to the strong support and considerable help from all sectors of society, is inadequate in terms of perspective and insight due to our limited research experience and tight deadline. Therefore, we welcome valuable opinions and suggestions from governmental departments, international organizations, research institutes, Internet corporations and social organizations across different sectors, home and abroad, to help us produce better reports in the future and contribute more wisdom and strength to world Internet development.

Chinese Academy of Cyberspace Studies (CACS)
October 2020

MIX
Papier aus verantwortungsvollen Quellen
Paper from responsible sources
FSC® C105338

If you have any concerns about our products,
you can contact us on
**ProductSafety@springernature.com**

In case Publisher is established outside the EU,
the EU authorized representative is:
**Springer Nature Customer Service Center GmbH**
**Europaplatz 3, 69115 Heidelberg, Germany**

Printed by Libri Plureos GmbH
in Hamburg, Germany